Water Conservation Handbook

Water Conservation Handbook

Edited by **Keith Wheatley**

New York

Published by Callisto Reference,
106 Park Avenue, Suite 200,
New York, NY 10016, USA
www.callistoreference.com

Water Conservation Handbook
Edited by Keith Wheatley

International Standard Book Number: 978-1-63239-606-8 (Hardback)

Printed in the United States of America.

Contents

Preface

The main aim of this book is to educate learners and enhance their research focus by presenting diverse topics covering this vast field. This is an advanced book which compiles significant studies by distinguished experts in the area of analysis. This book addresses successive solutions to the challenges arising in the area of application, along with it; the book provides scope for future developments.

The need for conservation of water is of utmost importance in the present day scenario. Water is an essential and basic human need for urban, industrial and agricultural usage. Though there is abundance in fresh water resources, its uneven distribution across the world has generated challenges for sustainable use of this resource. Water conservation refers to an efficient and optimal usage of water, and the protection of valuable water resources. This book lays emphasis on some commonly used tools and techniques like rainwater harvesting, water reuse and recycling, cooling water recycling, irrigation techniques like drip irrigation, agricultural management methods, groundwater management and water conservation incentives.

It was a great honour to edit this book, though there were challenges, as it involved a lot of communication and networking between me and the editorial team. However, the end result was this all-inclusive book covering diverse themes in the field.

Finally, it is important to acknowledge the efforts of the contributors for their excellent chapters, through which a wide variety of issues have been addressed. I would also like to thank my colleagues for their valuable feedback during the making of this book.

Editor

Analysis of Potable Water Savings Using Behavioural Models

Marcelo Marcel Cordova and Enedir Ghisi

Federal University of Santa Catarina, Department of Civil Engineering, Laboratory of Energy Efficiency in Buildings, Florianópolis – SC
Brazil

1. Introduction

The availability of drinking water in reasonable amounts is currently considered the most critical natural resource of the planet (United Nations Educational, Scientific and Cultural Organization [UNESCO], 2003). Studies show that systems of rainwater harvesting have been implemented in different regions such as Australia (Fewkes, 1999a; Marks et al., 2006), Brazil (Ghisi et al., 2009), China (Li & Gong, 2002; Yuan et al., 2003), Greece (Sazakli et al., 2007), India (Goel & Kumar, 2005; Pandey et al., 2006), Indonesia (Song et al., 2009), Iran (Fooladman & Sepaskhah, 2004), Ireland (Li et al., 2010), Jordan (Abdulla & Al-Shareef, 2009), Namibia (Sturm et al., 2009), Singapore (Appan, 1999), South Africa (Kahinda et al., 2007), Spain (Domènech & Saurí, 2011), Sweden (Villareal & Dixon, 2005), UK (Fewkes, 1999a), USA (Jones & Hunt, 2010), Taiwan (Chiu et al., 2009) and Zambia (Handia et al., 2003).

One of the most important steps in planning a system for rainwater harvesting is a method for determining the optimal capacity of the rainwater tank. It should be neither too large (due to high costs of construction and maintenance) nor too small (due to risk of rainwater demand not being met). This capacity can be chosen from economic analysis for different scenarios (Chiu et al., 2009) or from the potential savings of potable water for different tank sizes (Ghisi et al., 2009).

Several methodologies for the simulation of a system for rainwater harvesting have been proposed. The approaches commonly used are behavioural (Palla et al., 2011; Fewkes, 1999b; Imteaz et al., 2011; Ward et al., 2011; Zhou et al., 2010; Mitchell, 2007) and probabilistic (Basinger et al., 2010; Chang et al., 2011; Cowden et al., 2008; Su et al., 2009; Tsubo et al., 2005).

One advantage of the behavioural methods is that they can measure several variables of the system over time, such as volumes of consumed and overflowed rainwater, percentage of days in which rainwater demand is met (Ghisi et al., 2009), etc. The main disadvantage of these methods is that as the simulation is based on a mass balance equation, there is no guarantee of similar results when using different rainfall data from the same region (Basinger et al., 2010). This problem can be avoided, in part, with the use of long-term rainfall time series.

Probabilistic methods have the advantage of their robustness, for example, by using stochastic precipitation generators. A disadvantage of these methods is their portability. Several models adequately describe the rainfall process in one location but may not be satisfactory in another (Basinger et al., 2010).

A way of comparing different models for rainwater harvesting systems is by assessing their potential for potable water savings and optimal tank capacities.

The objective of this study is to compare the potential for potable water savings using three behavioural models for rainwater harvesting in buildings. The analysis is performed by varying rainwater demand, potable water demand, upper and lower tank capacities, catchment area and rainfall data.

Studies which consider behavioural models generally use either Yield After Spillage (YAS) or Yield Before Spillage (YBS) (Jenkins et al., 1978). This study aims to compare them with a software named Neptune (Ghisi et al., 2011). A method for determining the optimal tank capacity will also be presented based on the potential for potable water savings.

2. Methodology

Behavioural methods are based on mass balance equations. A simplified model is given by Eq. (1).

$$V(t) = Q(t) + V(t - 1) - Y(t) - O(t) \tag{1}$$

where V is the stored volume (litres), Q is the inflow (litres), Y is the rainwater supply (litres), and O is the overflow (litres).

The software named Neptune was used to perform the simulations. YAS and YBS methods were implemented only for simulations in this research, but they are not available to users.

Neptune requires the following data for simulation: daily rainfall time series (mm); catchment area (m²); number of residents; daily potable water demand (litres per capita/day); percentage of potable water that can replaced with rainwater; runoff coefficient; lower tank capacity; and upper tank capacity (if any).

For each day of the rainfall time series, Neptune estimates: the volume of rainwater that flows on the catchment surface area, the stored volume in the lower tank (at the beginning and end of the day), the overflow volume and the volume of rainwater consumed. If an upper tank is used, the volume stored in the upper tank and the volume of rainwater pumped from the lower to the upper tank are also estimated.

The volume of rainwater that flows on the catchment surface is estimated by using Eq. (2).

$$V_{catch}(t) = P(t) \cdot S \cdot C \tag{2}$$

where V_{catch} is the volume of rainwater that flows on the catchment surface (litres); P is the precipitation in day t (mm); S is the catchment surface area (m²); C is the runoff coefficient (non-dimensional, $0 < C \le 1$).

The methods Neptune, YAS and YBS differ in the way stored volumes are calculated and pumped. Details about them are shown as follows.

2.1 Neptune

The volume of rainwater stored in the lower tank at the beginning of a given day is calculated using Eq. (3).

$$V_{in\,low}(t) = min \begin{cases} V_{low\,tank} \\ V_{catch}(t) + V_{end\,low}(t - 1) \end{cases} \tag{3}$$

where $V_{in\ low}(t)$ is the volume of rainwater stored in the lower tank at the beginning of day t (litres); $V_{low\ tank}$ is the capacity of the lower tank (litres); $V_{catch}(t)$ is the volume of rainwater that flows on the catchment surface on day t (litres); $V_{end\ low}(t)$ is the volume of rainwater available in the lower tank at the end of the day (litres).

Next, the volume of rainwater that can be pumped to the upper tank is calculated by using Eq. (4).

$$V_{pump}(t) = min \begin{cases} V_{in\ low}(t) \\ V_{up\ tank} - V_{end\ up}(t-1) \end{cases} \tag{4}$$

where $V_{pump}(t)$ is the volume of rainwater pumped on day t (litres); $V_{in\ low}(t)$ is the volume of rainwater stored in the lower tank at the beginning of day t (litres); $V_{up\ tank}$ is the capacity of the upper tank (litres); $V_{end\ up}(t-1)$ is the volume of rainwater available in the upper tank at the end of the previous day (litres).

The volume of rainwater available in the lower tank at the end of a day is defined as the difference between the volume of rainwater in the beginning of the day and the volume that was pumped (Eq. (5)(4)).

$$V_{end\ low}(t) = V_{in\ low}(t) - V_{pump}(t) \tag{5}$$

where $V_{end\ low}(t)$ is the volume of rainwater available in the lower tank at the end of day t (litres); $V_{in\ low}(t)$ is the volume of rainwater stored in the lower tank at the beginning of day t (litres); $V_{pump}(t)$ is the volume of rainwater pumped on day t (litres).

The volume of rainwater available in the upper tank at the beginning of a given day (after pumping) is given by Eq. (6).

$$V_{in\ up}(t) = V_{end\ up}(t-1) + V_{pump}(t) \tag{6}$$

where $V_{in\ up}(t)$ is the volume of rainwater available in the upper tank at the beginning of day t (litres); $V_{end\ up}(t-1)$ is the volume of rainwater available in the upper tank at the end of the previous day (litres); $V_{pump}(t)$ is the volume of rainwater pumped on day t (litres).

The volume of rainwater consumed daily depends on rainwater demand and volume stored in the upper tank; it is calculated by using Eq. (7).

$$V_c(t) = min \begin{cases} D(t) \\ V_{in\ up}(t) \end{cases} \tag{7}$$

where $V_c(t)$ is the volume of rainwater consumed in day t (litres); $D(t)$ is the rainwater demand in day t (litres per capita/day); $V_{in\ up}(t)$ is the volume of rainwater available in the upper tank at the beginning of day t (litres).

The volume of rainwater available in the upper tank at the end of a given day is obtained by using Eq. (8).

$$V_{end\ up}(t) = V_{in\ up}(t) - V_c(t) \tag{8}$$

where $V_{end\ up}(t)$ is the volume of rainwater available in the upper tank at the end of day t (litres); $V_{in\ up}(t)$ is the volume of rainwater available in the upper tank at the beginning of day t (litres); $V_c(t)$ is the volume of rainwater consumed on day t (litres).

The potential for potable water savings results from the relationship between the total volume of rainwater consumed and the potable water demand over the period considered in the analysis, according to Eq. (9).

$$E_{pot} = 100 \cdot \sum_{t=1}^{T} \frac{V_c(t)}{D(t) \cdot N}$$

(9)

where E_{pot} is the potential for potable water savings (%); $V_c(t)$ is the volume of rainwater consumed on day t (litres); $D(t)$ is the rainwater demand on day t (litres per capita/day); N is the number of inhabitants; T is the period considered in the analysis (the same as the duration of the rainfall time series).

2.2 YAS

In the YAS method, the volume of rainwater collected will be consumed only in the next day. Thus, in systems where there is an upper and a lower tank, rainwater will be pumped at the beginning of the next day (Chiu & Liaw, 2008).

When considering the use of an upper tank, the difference between YAS and Neptune resides only in calculating the volume of rainwater pumped. It can be seen, in Eq. (10), that YAS method considers the volume stored in the tank at the previous day.

$$V_{pump}(t) = min \begin{cases} V_{in\ low}(t-1) \\ V_{up\ tank} - V_{end\ up}(t-1) \end{cases}$$

(10)

where $V_{pump}(t)$ is the volume of rainwater pumped on day t (litres); $V_{in\ low}(t-1)$ is the volume of rainwater stored in the lower tank at the beginning of the previous day (litres); $V_{up\ tank}$ is the capacity of the upper tank (litres); $V_{end\ up}(t-1)$ is the volume available in the upper tank at the end of the previous day (litres).

The other equations are identical to those presented for Neptune.

2.3 YBS

In Neptune and YAS methods, the available volume of rainwater at the end of a given day is estimated by using Eq. (8). Thus, it is possible to notice that the tank is never full at the end of the day, no matter the amount of rainwater available.

The main feature of the YBS method is the possibility that this gap does not exist. When using both upper and lower tanks, a way to fill the upper tank is pumping rainwater two times a day; the first time before or during consumption and the second one after consumption (usually at night).

For YBS method, the volume of rainwater stored in the lower tank at the beginning of day t is the same as that for Neptune and YAS, given by Eq. (3).

Thus, according to YBS method, the first volume of rainwater to be pumped is calculated by using Eq. (11).

$$V_{pump}(t) = min \begin{cases} V_{in\ low}(t) \\ V_{up\ tank} - V_{end\ up}(t-1) \end{cases}$$

(11)

where $V_{pump}(t)$ is the volume of rainwater pumped on day t (litres); $V_{in\ low}(t)$ is the volume of rainwater stored in the lower tank at the beginning of day t (litres); $V_{up\ tank}$ is the volume of the upper tank (litres); $V_{end\,up}(t-1)$ is the volume of rainwater available in the upper tank at the end of the previous day (litres).

The volume of rainwater available in the lower tank after the first pumping is given by Eq. (12).

$$V_{low\,aft\,pump}(t) = min \begin{cases} V_{low\,tank} \\ V_{end\,low}(t-1) + V_{catch}(t) - V_{pump}(t) \end{cases} \quad (12)$$

where $V_{low\,aft\,pump}(t)$ is the volume of rainwater available in the lower tank after the first pumping (litres); $V_{low\,tank}$ is the capacity of the lower tank (litres); $V_{end\,low}(t-1)$ is the volume of rainwater available in the lower tank at the end of the previous day (litres); $V_{catch}(t)$ is the volume of rainwater that flows on the catchment surface (litres); $V_{pump}(t)$ is the volume of rainwater pumped on day t (litres).

The volume of rainwater available in the upper tank after the first pumping is given by Eq. (13).

$$V_{in\,up}(t) = V_{end\,up}(t-1) + V_{pump}(t) \quad (13)$$

where $V_{in\,up}(t)$ is the volume of rainwater available in the upper tank after the first pumping (litres); $V_{end\,up}(t-1)$ is the volume of rainwater available in the upper tank at the end of the previous day (litres); $V_{pump}(t)$ is the volume of rainwater pumped on day t (litres).

The volume of rainwater consumed in a given day is calculated by using Eq. (14).

$$V_c(t) = min \begin{cases} D(t) \\ V_{in\,up}(t) \end{cases} \quad (14)$$

where $V_c(t)$ is the volume of rainwater consumed on day t (litres); $D(t)$ is the rainwater demand on day t (litres per capita/day); $V_{in\,up}(t)$ is the volume of rainwater available in the upper tank at the beginning of day t (litres).

After that consumption, the volume of rainwater available in the upper tank is calculated by using Eq. (15).

$$V_{up\,aft\,cons}(t) = V_{in\,up}(t) - V_c(t) \quad (15)$$

where $V_{up\,aft\,cons}(t)$ is the volume of rainwater available in the upper tank after consumption (litres); $V_{in\,up}(t)$ is the volume of rainwater available in the upper tank at the beginning of day t (litres); $V_c(t)$ is the volume of rainwater consumed on day t (litres).

The volume of rainwater available for the second pumping is given by Eq. (16).

$$V_{pump\,2}(t) = min \begin{cases} V_{low\,aft\,pump}(t) \\ V_{up\,tank} - V_{up\,aft\,cons}(t) \end{cases} \quad (16)$$

where $V_{pump\,2}(t)$ is the volume of rainwater available for the second pumping (litres); $V_{low\,aft\,pump}(t)$ is the volume of rainwater available in the lower tank after the first pumping (litres); $V_{up\,tank}$ is the capacity of the upper tank (litres); $V_{up\,aft\,cons}(t)$ is the volume of rainwater available in the upper tank after consumption (litres).

The volume of rainwater available in the upper and lower tanks at the end of a given day are given by Eqs. (17) and (18), respectively.

$$V_{end\,up}(t) = min \begin{cases} V_{up\,tank} \\ V_{up\,aft\,cons}(t) + V_{pump\,2}(t) \end{cases} \qquad (17)$$

where $V_{end\,up}(t)$ is the volume of rainwater available in the upper tank at the end of day t (litres); $V_{up\,tank}$ is the capacity of the upper tank (litres); $V_{up\,aft\,cons}(t)$ is the volume of rainwater available in the upper tank after consumption (litres); $V_{pump\,2}(t)$ is the volume of rainwater available for the second pumping (litres).

$$V_{end\,low}(t) = V_{low\,aft\,pump}(t) - V_{pump\,2}(t) \qquad (18)$$

where $V_{end\,low}(t)$ is the volume of rainwater available in the lower tank at the end of the day (litres); $V_{low\,aft\,pump}(t)$ is the volume of rainwater available in the lower tank after the first pumping (litres); $V_{pump\,2}(t)$ is the volume of rainwater available for the second pumping (litres).

2.4 Computer simulations

In order to compare Neptune, YAS and YBS, computer simulations were carried out for different cases. Table 1 shows the parameters considered for the simulations.

Parameter	Case 1 – Low rainwater demand	Case 2 – Medium rainwater demand	Case 3 – High rainwater demand
Catchment surface area (m²)	100	200	300
Potable water demand (litres per capita/day)	100	200	300
Number of inhabitants per house	3	4	5
Percentage of potable water that can be replaced with rainwater (%)	30	40	50
Total rainwater demand (litres/day per house)	90	320	750
Capacity of the upper tank (litres)	90	320	750

Table 1. Simulation parameters for low, medium and high rainwater demand for Santana do Ipanema, Florianópolis and Santos.

In all three cases a runoff coefficient of 0.8 was taken into account, i.e., 20% of rainwater is discarded due to dirt on the roof, gutters, etc. The capacity of the upper tank is given by the daily rainwater demand. It is calculated by using Eq. (19).

$$V_{up\,tank} = D_{pot} \cdot N_{inh} \cdot P_{subst} \qquad (19)$$

where $V_{up\,tank}$ is the capacity of the upper tank (litres); D_{pot} is the potable water demand (litres); N_{inh} is the number of inhabitants; P_{subst} is the percentage of potable water that can be replaced with rainwater.

Three cities with different rainfall patterns were considered in the simulations: Santana do Ipanema, Florianópolis and Santos. The monthly average rainfall for the three cities are shown in Figure 1, Figure 2 and Figure 3, respectively.

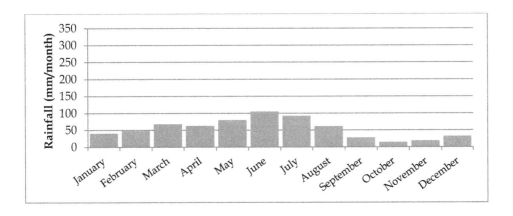

Fig. 1. Monthly average rainfall in Santana do Ipanema over 1979-2010.

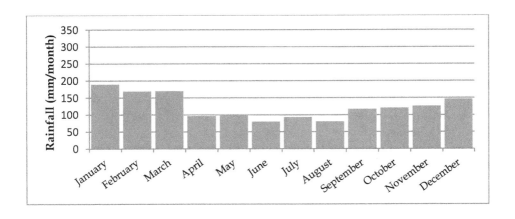

Fig. 2. Monthly average rainfall in Florianópolis over 1949-1998.

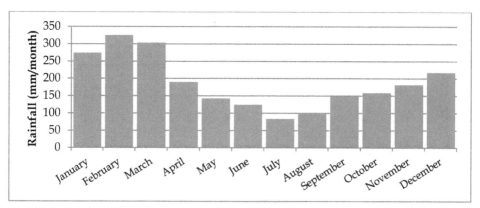

Fig. 3. Monthly average rainfall in Santos over 1910-1996.

The annual average rainfall for the three cities are: Santana do Ipanema – 652 mm; Florianópolis – 1486 mm; Santos – 2252 mm.

For the simulations, the last 10 years of daily rainfall data were used for each city. Data from 2001-2010 were used for Santanan do Ipanema; from 1989-1998 for Florianópolis , and from 1987-1996 for Santos.

2.5 Optimal capacity for the lower tank

To calculate the ideal capacity for the lower tank, simulations were performed for tank capacities ranging from 0 to 10,000 litres, at interval of 250 litres. Then graphs of potential for potable water savings as a function of tank capacities were drawn. For each two points in the graph, the difference between potable water savings was estimated by using Eq. (20).

$$\Delta_i = \frac{E_{pot}(i) - E_{pot}(i-1)}{V_{low\ tank}(i) - V_{low\ tank}(i-1)} \tag{20}$$

where Δ_i is difference between potable water savings ($\%/m^3$); E_{pot} is the potential for potable water savings (%); $V_{low\ tank}$ is the lower tank capacity (m^3).

Eq. (20) represents the resulting increase in E_{pot} for a given increase in V_{inf}. As "%/litre" usually results in very small values, the tank capacities are expressed in m^3.

The tank capacity chosen as optimal is the one in which $\Delta_i \leq 1\%/m^3$. This means that, for that interval, an increase of 1 m^3 in the capacity of the lower tank results in an increase less or equal to 1% in the potential for potable water savings.

This ensures that the tank capacity will not be too small (such that the rainwater demand will not be met) or too large (such that the tank will not be filled for most of the time).

3. Results

In this section, results for the three cases and three cities are shown. The optimal capacities for the lower tank are determined for YAS, YBS and Neptune.

It will be seen that the potential for potable water savings, in %, obtained with Neptune is always greater than YBS and smaller than YAS. Thus, to compare results for a given capacity, the reference will be that estimated by Neptune.

3.1 Low rainwater demand

The simulation for Santana do Ipanema gives the results shown in Figure 4.

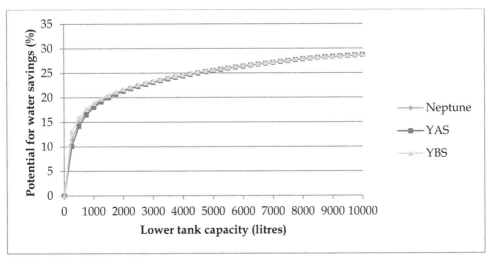

Fig. 4. Potential for potable water savings for Santana do Ipanema, with low rainwater demand.

Due to low rainfall, even with a low rainwater demand (90 litres/day), it can be seen that the maximum percentage of rainwater demand, 30%, is not reached within the range of tank capacities simulated.

The ideal capacities for the lower tanks are: Neptune – 4500 litres; YAS – 4750 litres; YBS – 4500 litres. The potential for potable water savings are, respectively, 25.15%, 25.31% and 25.24%.

Considering a tank capacity of 4500 litres, additional results are obtained (Table 2).

Parameter	Neptune	YAS	YBS
Volume of rainwater overflowed (litres)	26,826	26,943	26,727
Daily average of volume overflowed (litres/day)	7.4	7.4	7.3
Volume of rainwater consumed (litres)	275,554	274,384	276,570
Daily average of volume consumed (litres/day)	75.9	75.2	75.8
Percentage of days that rainwater demand is completely met	83.19	82.83	83.54
Percentage of days that rainwater demand is partially met	1.23	1.23	1.15
Percentage of days that rainwater demand is not met	15.58	15.94	15.31

Table 2. Results for Santana do Ipanema for low rainwater demand and a lower tank capacity of 4500 litres.

The difference between average rainwater consumption for Neptune and YAS is 0.32 litres/day, which is equivalent to 0.36% of daily rainwater demand. Similarly, the difference between YBS and Neptune is 0.28 litres/day, which corresponds to 0.31% of daily rainwater demand.

For Florianópolis, the potential for potable water savings as a function of the volume of lower tank is presented in Figure 5.

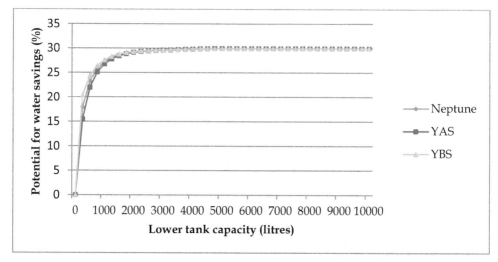

Fig. 5. Potential for potable water savings for Florianópolis, with low rainwater demand.

For Florianópolis, which has greater rainfall than Santana do Ipanema, one sees that, with tank capacity around 3000 litres the maximum potential for water savings is reached.

The ideal capacities for the lower tanks are: Neptune – 2000 litres; YAS – 2000 litres; YBS – 1750 litres. The potential for potable water savings are, respectively, 29.24%, 29.15% and 29.08%.

Table 3 presents additional results for the three methods using a lower tank of 2000 litres.

Parameter	Neptune	YAS	YBS
Volume of rainwater overflowed (litres)	103,548	103,649	103,473
Daily average of volume overflowed (litres/day)	31.2	31.3	31.2
Volume of rainwater consumed (litres)	290,840	289,924	291,432
Daily average of volume consumed (litres/day)	87.7	87.5	87.9
Percentage of days that rainwater demand is completely met	97.20	96.83	97.44
Percentage of days that rainwater demand is partially met	0.51	0.57	0.36
Percentage of days that rainwater demand is not met	2.29	2.60	2.20

Table 3. Results for Florianópolis for low rainwater demand and a lower tank of 2000 litres.

The difference between average rainwater consumption for Neptune and YAS is 0.28 litres/day, which is equivalent to 0.31% of daily rainwater demand. Similarly, the difference between YBS and Neptune is 0.18 litres/day, which corresponds to 0.20% of daily rainwater demand.

The potential for potable water savings for Santos is presented in Figure 6.

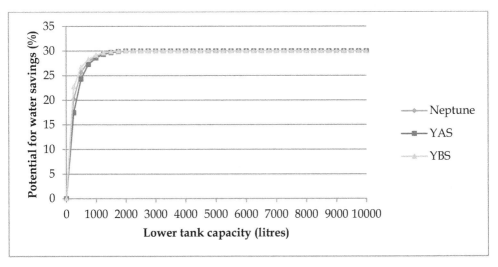

Fig. 6. Potential for potable water savings for Santos, with low demand of rainwater.

In this case, the maximum potential for potable water savings is reached for a lower tank capacity of about 2000 litres.

The ideal capacities for the lower tanks are: Neptune – 1500 litres; YAS – 1500 litres; YBS – 1500 litres. The potential for potable water savings are, respectively, 29.76%, 29.67% and 29.84%.

Table 4 presents additional results for the three methods using a lower tank of 1500 litres.

Parameter	Neptune	YAS	YBS
Volume of rainwater overflowed (litres)	250,974	251,075	250,924
Daily average of volume overflowed (litres/day)	68.8	68.8	67.8
Volume of rainwater consumed (litres)	321460	320460	322228
Daily average of volume consumed (litres/day)	88.1	87.8	88.3
Percentage of days that rainwater demand is completely met	99.06	98.72	99.33
Percentage of days that rainwater demand is partially met	0.25	0.31	0.20
Percentage of days that rainwater demand is not met	0.69	0.97	0.47

Table 4. Results for Santos for low rainwater demand and a lower tank of 1500 litres.

The difference between average rainwater consumption for Neptune and YAS is 0.27 litres/day, which is equivalent to 0.27% of daily rainwater demand. Similarly, the difference between YBS and Neptune is 0.21 litres/day, which corresponds to 0.23% of daily rainwater demand.

3.2 Medium rainwater demand

Considering a daily rainwater demand of 320 litres and a catchment surface of 200 m², the shape of the curves on the graphs remain the same, with an asymptotic tendency.

For Santana do Ipanema, the maximum potential for potable water savings (40%) cannot be reached due to small amounts of rainfall. The ideal capacity for the lower tank with method Neptune is 5000 litres. YAS estimated a capacity 250 litres bigger, while YBS estimated a capacity 250 litres smaller. The potential for potable water savings are, respectively, 23.29%, 23.26% and 23.36%. With a lower tank with capacity of 5000 litres, the difference between average rainwater consumption for Neptune and YAS is equivalent to 0.78% of daily rainwater demand. Similarly, the difference between YBS and Neptune corresponds to 0.73% of daily rainwater demand.

The ideal capacities for the lower tank using Neptune and YAS were the same as those estimated for Santana do Ipanema. YBS had an optimal capacity of 4500 litres. However, due to higher rainfall the potential for potable water savings are, respectively, 36.34%, 36.27% and 36.17%. With a lower tank capacity of 5000 litres, the difference between average rainwater consumption for Neptune and YAS corresponds to 0.82% of daily rainwater demand. Similarly, the difference between YBS and Neptune is equivalent to 0.71% of daily rainwater demand.

As an example, Figure 7 shows the potential for potable water savings as a function of the lower tank capacity for Santos.

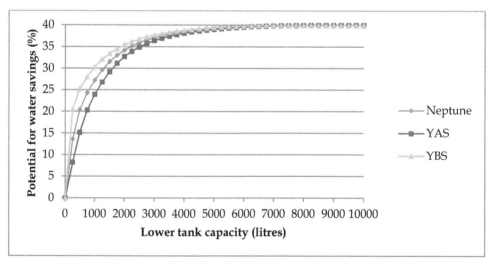

Fig. 7. Potential for potable water savings for Santos, with medium rainwater demand.

Santos, which has higher rainfall than Santana do Ipanema and Florianópolis, can reach the maximum potential for potable water savings, with a tank capacity of about 7000 litres. The ideal capacities, however, are considerably smaller. The estimated capacities for Neptune, YAS and YBS were, respectively, 4000 litres, 4250 litres and 3750 litres. For these lower tanks, the potential for potable water savings are 38.49%, 38.42% and 38.56%. With a lower tank capacity of 4000 litres, the difference between average rainwater consumption for Neptune and YAS is equivalent to 0.89% of daily rainwater demand. Likewise, the difference between YBS and Neptune corresponds to 0.68% of daily rainwater demand.

3.3 High rainwater demand
The third case considers a higher rainwater demand, i.e., 750 litres/day. The catchment surface is also larger, i.e., 300 m².
For Santana do Ipanema, which has the lowest rainfall, the simulation gives the results shown in Figure 8.

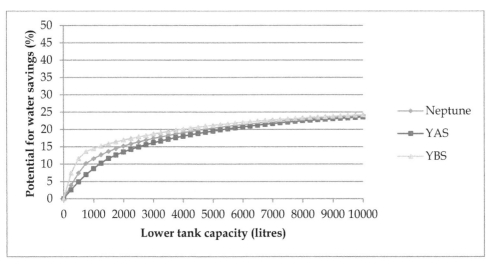

Fig. 8. Potential for potable water savings for Santana do Ipanema, with high rainwater demand.

Due to low rainfall in Santana do Ipanema, and the high rainwater demand, the highest potential for potable water savings obtained in the interval 0-10000 litres is less than 25%. Differences in the lower tank capacity are greater than the ones obtained in the previous sections. The ideal capacities for Neptune, YAS and YBS are 5500 litres, 6250 litres and 4750 litres, respectively. The potential for potable water savings, on the other hand, are very similar: respectively 20.90%, 20.97% and 20.90%. Considering a lower tank capacity of 5500 litres, the difference between average rainwater consumption for Neptune and YAS corresponds to 1.46% of daily rainwater demand. Similarly, the difference between YBS and Neptune is equivalent to 1.43% of daily rainwater demand.
For Florianópolis, a potential for potable water savings of 40% is the most that can be obtained in the interval 0-10000 litres, due to the high rainwater demand. The ideal capacities for the lower tanks are: Neptune – 8250 litres; YAS – 9000 litres; YBS – 7500 litres. The potential for potable water savings, however, are almost equal: 39.63%, 39.65% and

39.63%, respectively. The biggest difference in the average rainwater consumption occurs between Neptune and YAS, and is equivalent to 1.50% of daily rainwater demand.

Because of higher amounts of rainfall, lower tank capacities estimated for Santos are smaller than those obtained for Florianópolis. For Neptune, it is 7750 litres. YAS and YBS estimated volumes of 8500 litres and 7000 litres, respectively. The potential for potable water savings are, respectively, 46.10%, 46.11% and 46.79%. With a lower tank capacity of 7750 litres, the difference between average rainwater consumption for Neptune and YAS is equivalent to 1.65% of daily rainwater demand. Similarly, the difference between YBS and Neptune corresponds to 1.35% of daily rainwater demand.

As noted in the previous sections, the differences between methods are very small compared to the daily rainwater demand.

4. Conclusions

Three behavioural models for rainwater harvesting analysis were investigated. Two rainwater tanks were considered, i.e., a lower and an upper one, so that the water is pumped from the lower to the upper tank.

A methodology for determining the optimum lower tank capacity was presented, based on variations in the potential for potable water savings as a function of the tank capacity. Results showed that the method estimates a capacity for the lower tank that is not too small so as to allow for a great amount of rainwater to be wasted; and neither too large so as to allow for the increase in construction and maintaining costs to surpass the increase in potential for potable water savings.

Simulations were performed for three rainwater demands and three cities. Results showed that, due to the modelling, the YAS method always estimates the smallest potential for potable water savings, followed by Neptune and YBS, respectively. It was also found that the differences between the methods increase as increases the rainwater demand.

Despite the potential for potable water savings obtained with YBS being slightly higher than the other two methods, one should take into account that two pumpings per day can occur; and this causes an increase in system maintenance and energy costs.

The greatest difference of daily average rainwater consumed obtained between Neptune and YAS was 1.65%. Similarly, the greatest difference between Neptune and YBS was 1.35%. Thus, it can be concluded that, for practical purposes, the methods are equivalent.

5. References

Abdulla, F. A. & Al-Shareef, A. W. (2009). Roof rainwater harvesting systems for household water supply in Jordan. *Desalination*, n. 243, p. 195-207.

Appan, A. (1999). A dual-mode system for harnessing roofwater for non-potable uses. *Urban Water*, n. 1, p. 317-321.

Basinger, M.; Montalto, F. & Lall, U. (2010). A rainwater harvesting system reliability model based on nonparametric stochastic rainfall generator. *Journal of Hydrology*, n. 392, p. 105-118.

Chang, N.; Rivera, B. J. & Wanielista, M. P. (2011). Optimal design for water conservation and energy savings using green roofs in a green building under mixed uncertainties. *Journal of Cleaner Production*, n. 19, p. 1180-1188.

Chiu, Y. & Liaw, C. (2008). Designing rainwater harvesting systems for large-scale potable water saving using spatial information system. *Lecture Notes in Computer Science*, v. 5236, p. 653-66.

Chiu, Y.; Liaw, C. & Chen, L. (2009). Optimizing rainwater harvesting systems as an innovative approach to saving energy in hilly communities. *Renewable Energy*, n. 34, p. 492-498.

Cowden, J. R.; Watkins Jr., D. W. & Mihelcic, J. R. (2008). Stochastic rainfall modeling in West Africa: Parsimonious approaches for domestic rainwater harvesting assessment. *Journal of Hydrology*, n. 361, p. 64-77.

Domènech, L &; Saurí, D. (2011). A comparative appraisal of the use of rainwater harvesting in single and multi-family buildings of the Metropolitan Area of Barcelona (Spain): social experience, drinking water savings and economic costs. *Journal of Cleaner Production*, n. 19, p. 598-608.

Fewkes, A. (1999a). The use of rainwater for WC flushing: the field testing of a collection system. *Building and Environment*, n. 34, p. 765-772.

Fewkes, A. (1999b). Modelling the performance of rainwater collection systems: towards a generalized approach. *Urban Water*, n. 1, p. 323-333.

Fooladman, H. R. & Sepaskhah, A. R. (2004). Economic analysis for the production of four grape cultivars using microcatchment water harvesting systems in Iran. *Journal of Arid Environments*, v. 58, p. 525-533.

Ghisi, E. Tavares, D. F. & Rocha, V. L. (2009). Rainwater harvesting in petrol stations in Brasília: Potential for potable water savings and investment feasibility analysis. *Resources, Conservation and Recycling*, v. 54, p. 79-85.

Ghisi, E.; Cordova. M. M. & Rocha, N. L. (2011). *Neptune 3.0*. Computer programme. Federal University of Santa Catarina, Department of Civil Engineering. Available in: http://www.labeee.ufsc.br.

Goel, A. K. & Kumar, R. (2005). Economic analysis of water harvesting in a mountainous watershed in India. *Agricultural Water Management*, v. 71, p. 257-266.

Handia, L.; Tembo, J. M. & Mwiindwa, C. (2003). Potential of Rainwater harvesting in urban Zambia. *Physics and Chemistry of the Earth*, v. 28, p. 893-896.

Imteaz, M. A.; Shanableh, A.; Rahman, A. & Ahsan, A. (2011). Optimisation of rainwater tank design from large roofs: A case study in Melbourne, Australia. *Resources, Conservation and Recycling*, Article in press.

Jenkins, D.; Pearson, F.; Moore, E.; Sun, J. K. & Valentine, R. (1978). *Feasibility of rainwater collection systems in California*. Californian Water Resources Centre, University of California, USA.

Jones, M. P. & Hunt, W. F. (2010). Performance of rainwater harvesting systems in the south eastern United States. *Resources, Conservation and Recycling*, v. 54, p. 623-629.

Kahinda, J. M.; Taigbenu, A. E. & Boroto, J. R. (2007). Domestic Rainwater harvesting to improve water supply in rural South Africa. *Physics and Chemistry of the Earth*, v. 32, p. 1050-1057.

Li, X. & Gong, J. (2002). Compacted microcatchments with local earth materials for rainwater harvesting in the semiarid region of China. *Journal of Hydrology*, v. 257, p. 134-144.

Li, Z.; Boyle, F. & Reynolds, A. (2010). Rainwater harvesting and greywater treatment systems for domestic application in Ireland. *Desalination*, v. 260, p. 1-8.

Marks, R.; Clark, R.; Rooke, E. & Berzins, A. (2006). Meadows, South Australia: development through integration of local water resources. *Desalination*, v. 188, p. 149-161.

Mitchell, V. G. (2007). How important is the selection of computational analysis method to the accuracy of rainwater tank behaviour modelling. *Hydrological Processes*, v. 21, p. 2850-2861.

Palla, A.; Gnecco, I. & Lanza, L. G. (2011). Non-dimensional design parameters and performance assessment of Rainwater harvesting systems. *Journal of Hydrology*, v. 401, p. 65-76.

Pandey, P. K.; Panda, S. N. & Panigrahi, B. (2006). Sizing on-farm reservoirs for crop-fish integration in rainfed farming systems in Eastern India. Biosystems Engineering, v. 93, p. 475-489.

Sazakli, E.; Alexopoulos, A. & Leotsinidis, M. (2007). Rainwater harvesting, quality assessment and utilization in Kefalonia Island, Greece. *Water Research*, v. 41, p. 2039-2047.

Song, J.; Han, M.; Kim, T. & Song, J. (2009). Rainwater harvesting as a sustainable water supply option in Banda Aceh. *Desalination*, v. 248, p. 233-240.

Sturm, M.; Zimmermann, M.; Schütz, K.; Urban, W. & Hartung, H. (2009). Rainwater harvesting as an alternative water resource in rural sites in central northern Namibia. *Physics and Chemistry of the Earth*, v. 34, p. 776-785.

Su, M.; Lin, C.; Chang, L.; Kang, J. & Lin, Mei. (2009). A probabilistic approach to rainwater harvesting systems design and evaluation. *Resources, Conservation and Recycling*, v. 53, p. 393-399.

Tsubo, M.; Walker, S. & Hensley, M. (2005). Quantifying risk for water harvesting under semi-arid conditions: Part I. Rainfall intensity generation. *Agricultural Water Management*, v. 76, p. 77-93.

United Nations Educational, Scientific and Cultural Organization (UNESCO). (2003). *The 1st UN World Water Development Report:* Water for People, Water for Life. Available in: <http://www.unesco.org/water/wwap/wwdr/wwdr1/table_contents/index.sht ml>.

Villareal, E. L. & Dixon, A. (2005). Analysis of a rainwater collection system for domestic water supply in Ringdansen, Norrköping, Sweden. *Building and Environment*, v. 40, p. 1174-1184.

Ward, S.; Memon, A. & Butler, D. (2011). Rainwater harvesting: model-based design evaluation. *Water Science and Technology*, v. 61, n. 1, p. 85-96.

Yuan, T.; Fengmin, L. & Puhai, L. (2003). Economic analysis of rainwater harvesting and irrigation methods, with an example from China. *Agricultural Water Management*, v. 60, p. 21-226.

Zhou, Y.; Shao, W. & Zhang, T. (2010). Analysis of a Rainwater harvesting system for domestic water supply in Zhoushan, China. *Journal of Zhejian University*, v. 11, n. 5, p. 342-348.

2

Review of Water-Harvesting Techniques to Benefit Forage Growth and Livestock on Arid and Semiarid Rangelands

Albert Rango and Kris Havstad

USDA-ARS Jornada Experimental Range, Las Cruces, New Mexico,
USA

1. Introduction

The term water harvesting means the concentration, collection, and distribution of water that would naturally exit a landscape through other processes (runoff, evaporation). Although very simple in concept and ancient in its history of application, it is surprising that this traditional water management approach is not more commonly implemented. When utilized, water harvesting is normally found in irrigated agriculture and domestic water supply applications, usually in less developed and impoverished regions of the globe. It has been found (Boers & Ben-Asher, 1982) that literature for water harvesting applied to crop production was sparser than expected. There have been applications on rangeland, particularly for desired management effects such as enhanced forage growth, landscape level distribution of livestock water supply, and rehabilitation of deteriorated or degraded resource conditions. Although applications in irrigated agriculture and domestic water supply are similar, this chapter focuses on the documented rangeland water harvesting approaches, which are even less common than those applications for crop production. (e.g., Frasier & Myers, 1983, Hudson, 1987, Critchley, et al., 1991, and Renner & Frazier, 1995).

The percentage of the world's total land surface area occupied by rangeland is between 40% to 70% depending on the definition used by the author (Branson, et al., 1981; Heady & Child 1994; and Holechek et al., 1995). Approximately 80% of all the world's rangeland is found in arid and semiarid regions (Branson et al., 1981), of which, the rangelands in the southwestern U.S. are good examples. The Jornada Experimental Range (Jornada) in south central New Mexico is representative of both the southwestern U.S. and the world's arid to semiarid rangeland and is a long-term ecological research site that has produced almost a century of important rangeland research knowledge (Havstad et al., 2006).

There are a number of unanswered questions that will be addressed in this chapter. Specifically, why haven't water harvesting techniques been used more frequently in arid and semiarid rangelands, and where they have been used, what problems have been encountered and what gaps in our knowledge still exist? Briefly, water harvesting has been used on rangelands but the documentation of the results are widely scattered. No insurmountable problems have been encountered, but a synthesis of existing results in one chapter should provide easier access to the existing literature for informed decisions on where and how to employ various water harvesting approaches. The authors have reviewed and assembled key water harvesting documentation which indicates that the techniques are easily used with the most effective approach for enhancing soil moisture and forage growth coming from constructing shallow water ponding dikes across known overland flow paths. For livestock watering, the construction of dirt stock tanks in established water channels provides valuable water sources for domestic animals as well as wildlife resulting in a complementary source of water to well water that may otherwise be wasted through the normal rainfall – evaporation cycle.

The authors have examined the positive and negative aspects of the water harvesting results. Surprisingly, there are few noted disadvantages. One disadvantage is a lack of evidence that shallow water ponding dikes have a capability to be self propagating in regards to vegetation growth. One distinct advantage is that simple traditional treatments are easy and inexpensive to install. Interestingly, historical and ancient methods are timeless. Infiltration of water into rangeland soils results in increased soil moisture and resulting associated forage growth. Still, future research is required over areas larger than those documented in the literature to see if results will vary based on differences in spatial scale. There will be a challenge in doing treatments as well as conducting measurements of soil moisture and forage over much larger areas than those used in the past. However, research at appropriate spatial scales should lead to more comprehensive recommendations of how to proceed in water harvesting in arid and semiarid rangelands around the world.

2. Historical applications

2.1 Ancient evidence

Investigators have found evidence in Jordan that water harvesting structures were constructed over 9,000 years ago and in Southern Mesopotamia over 6,500 years ago (Bruins et al., 1986). Water harvesting structures used by the Phoenicians in the Negev Desert were found to date back 3,000 – 4,000 years (Lowdermilk, 1960). Water collection and irrigation structures in southern Mexico have survived in excellent condition for about 3,000 years (Caran & Neely, 2006). Water collecting structures were also found in the Negev Desert dating back at least 2,700 years and probably longer (Evenari et al., 1982). Water harvesting for irrigation has been practiced in the desert areas of Arizona and northwest New Mexico for at least the last 1,000 years (Zaunderer & Hutchinson, 1988).

The rainwater harvesting approaches cited as used in the Negev Desert include terraces in wadis that are still under cultivation by local Bedouins and water harvesting farms reconstructed as part of an experiment by researchers at local universities (Evenari et al., 1982). Figure 1 is an aerial photo showing a farm unit near Shivta in the Negev desert that features terraces in the wadis that slow water flow (Evenari et al., 1982). This allows infiltration and an increase in soil moisture which enhances the success of cultivation behind the terraces. To increase the volume of water available for farming, stone-lined conduits from the surrounding hillsides collect and rapidly transmit rainfall runoff to the cultivated area.

Figure 2 is a schematic of a water spreading system illustrating floodwaters being delivered to a sequence of water ponding dikes that have historically been used on rangelands in the Middle East (Prinz & Malik, 2002, as adapted from French & Hussain, 1964). These types of water spreaders are typical of those used in arid regions around the world. However, as reported in Fedelibus & Bainbridge (1995), "like many great solutions to environmental problems, rainfall catchments" (or water harvesting methods) "are a reinterpretation of ancient techniques developed in the Middle East and Americas, but forgotten by modern science and technology."

2.2 Recent History

The availability of relatively inexpensive labor in the period 1934-1942 through Civilian Conservation Corps (CCC) personnel working at the direction of U.S. Government scientists produced a large number of land treatment measures throughout the western U. S. drylands. Peterson & Branson (1962) report that 899 water conservation structures established by the CCC were located and appraised in 1949 and 1961 in the Upper Gila and Mimbres River watersheds in Arizona and New Mexico. The effectiveness of the treatments were assessed in terms of vegetation improvement, longevity, and quantities of sediment retained by the structures. More than half of the structures were breached by water within several years after construction and were not functioning as planned. However, the most effective water applications were where earthen dikes were not breached and water was able to reach the spreader system, which resulted in vegetation improvement even in the driest areas of the region.

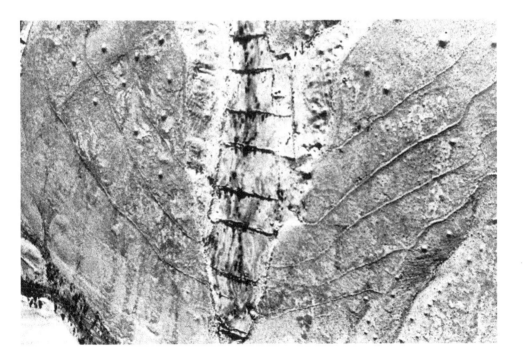

Fig. 1. Aerial photograph of a farm unit near Shivta in the Negev Desert. A terraced wadi and stone conduits leading runoff from hillsides to terraces are visible (after Evenari et al., 1982)

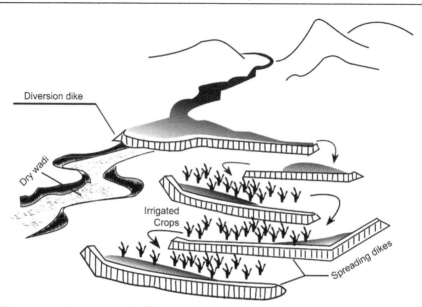

Fig. 2. Water spreading system in Pakistan to divert excess flood water ponding dikes (after Prinz & Malik, 2002, as adapted from French & Hussain, 1964).

Another study (Miller et al., 1969) of water spreader effectiveness found that the response of forage vegetation was dependent on rainfall characteristics, runoff production, and drainage of water detained in ponds behind dikes. If a site received less than 200 mm annual precipitation or less than 100-130 mm during the growing season, it would typically not produce enough runoff to justify installation of a water spreader (Bennett, 1939). Results produced by Valentine (1947), Hubbell & Gardner (1950), Hubbard & Smoliak (1953), Branson (1956), Houston (1960), and Hadley & McQueen (1961) showed increases in yield of forage grasses from small to large amounts (Miller et al., 1969). Forage production occurred only on those sites that received a minimum of at least one flooding event per year. The amount of soil moisture in the soil profile had more influence than soil texture on forage produced. More forage was also produced when ponded water could drain completely from the soil surface between rainfall events.

Similar work has also been done in other arid and semiarid regions of the world. As an example, Cunningham et al. (1974) have reported on the use of water ponding dikes to reclaim extensive bare soil areas (scalds) in Australia. This water ponding approach yielded almost double the amount of forage obtained from nearby non-scald areas with the same soil type. Scalds are formed through a combination of wind and water erosion removing surface soil to expose the subsoil which subsequently becomes very impervious (Warren, 1965; Cunningham et al., 1974). Soil berms of 30-45 cm high were constructed using a road grader that allowed ponding of surface runoff that was trapped behind a berm after a rainfall event.

Some of the most recent water ponding dikes constructed in the United States were on the Jornada to evaluate the efficiency of the shallow structures to increase forage. Twenty-five dikes in four separate areas of the Jornada were installed between 1975-1981 (Rango et al., 2006). These types of dikes can be constructed with a tractor and mold board plow or with a road grader as in this experiment (See Figure 3).

a
b
c

Fig. 3. Development of water ponding dikes at the Jornada showing a) 15cm dikes at the Ace Tank during construction with a road grader in 1975; b) 7.5cm dikes at Taylor Well during construction (accomplished using a mold board plow); and c) water ponding behind a 7.5cm dike at Taylor Well.

The height of the dikes ranged from a low of 7.5 cm to a high of 30 cm (See Table 1). The orientation of the dikes are typically perpendicular to the general direction of the overland flow. Additionally, multiple dikes are often arranged so that dikes downslope can catch any overflow from the upslope dikes. A crescent shape dike was usually employed to gather the water into a shallow pond. The resulting pattern of the dikes and vegetation growth approximate the pattern of natural banded vegetation which serves a similar function, namely, slowing down overland flow over bare areas, allowing the water to infiltrate into the soil moisture reservoir, and a resulting increase in vegetation growth (Tongway et al., 2001).

Location	Date Installed	Height	Number of Dikes	Soil Texture
Taylor Well	1975	7.5 cm	5	Fine
Ace Tank	1975	15 cm	5	Fine
Brown Tank	1978	15 cm	3	Fine
Dona Ana Exclosure	1981	30 cm	12	Medium-Coarse

Table 1. Attributes of water ponding dikes established on the Jornada Experimental Range in southcentral New Mexico.

Design criteria used when installing the water ponding dikes at the Jornada are specified by Tromble (1983). These criteria varied because of the characteristics of the individual site. Dikes were installed on fine to medium textured soils where the soil sealed rapidly during rainfall events thereby producing surface runoff. Furthermore, the dikes were placed in areas of "wasteland" supporting little or no vegetation (similar to the "scalds" in Australia). Dikes were placed starting at the highest place on the slope and working downslope. The direction of water flow from one dike to the next was regulated by locating one end of the dike higher than the other end so that water flowed out the lower end of the dike once it filled (Tromble, 1983). The distance between dikes was a function of the slope and expected water ponding depth. Enough distance was left between dikes to provide a source area for surface runoff water. Usually a 1:1 or 2:1 ratio of water runoff area to ponding area is satisfactory on a scald area (Tromble, 1983), but other authors cite values of 5:1 to 25:1 depending upon local conditions (Vallentine, 1989).

Compared to the shallow water ponding dikes, much deeper ponds or lakes are also necessary in rangelands for watering livestock as well as for providing fringe benefits to wildlife. Placement of these water sources depends not only upon water availability, but also on animal numbers, available forage, soil properties, and abiotic factors. Many approaches to providing water have been tested on rangelands. The two most usual developed water sources are from drilled, deep water wells and earthen dams that cause the formation of stock ponds. Early in the 20th century, it was recognized that competition to use reliable well water would become intense and that earthen dams to capture surface runoff would help reduce the pressure on well water utilization (Talbot, 1926). Simple, small earthen dams placed across drainage ways can provide open water storage during the rainy season and sometimes year round. This ancient technique has been used ubiquitously in arid regions and continues to be used throughout the world today. The ponds formed behind these earthen dams are, obviously, much deeper than behind shallow water ponding dikes.

In times of drought, stock ponds can dry up and the more reliable water source from deep wells is commonly relied upon. But for large portions of the year, especially in average to wet years, it is best to restrict deep well pumping to conserve the limited groundwater resources. In the same general water harvesting family that includes water ponding dikes and water spreaders, livestock water supply schemes can be very simple and are generally referred to as dirt tanks, stock tanks, or stock ponds.

In sparsely settled rangeland, especially in arid and semiarid regions where vegetation cover is limited, arguments for expensive solutions to make water available are usually difficult to justify. Reasons for deciding on earthen dam stock ponds are numerous, but the primary reason is that they are affordable to construct. Stock tanks are especially suitable in closed drainage, arid basins where flow in channels never leaves the basin, except by infiltrating into the stream channel bed and subsequently evaporating. By concentrating the flow behind earthen dams, the surface water area is small, thereby reducing both infiltration and evaporation. When surface soils in a watershed (and also used in construction) have a high percentage of clay and silty loams, infiltration rates tend to be low and surface runoff is increased. The use of stock ponds as the primary source of surface water reduces the use of well water with a more expensive infrastructure required for establishment. In most arid regions today, the natural recharge by precipitation is unable to provide replenishment of ground water because of an increasing deficit between recharge and pumping, both historic and present day (Giordano, 2009). Also, simple earthen dams and ponds last a long time with minimal maintenance. At the Jornada 77 stock ponds exist over the 783 km^2 area, most of which were constructed by Civil Conservation Corps labor in the 1930s (see Figures 4 and 5).

Fig. 4. Construction of stock water storage tanks at the Jornada Experimental Range using three-horse Fresno teams on Big Meadows tank by the Civilian Conservation Corps (CCC) in April 1934.

Fig. 5. Use of mechanized earth moving equipment during construction of Big Meadows tank at the Jornada Experimental Range in 1934. New equipment and the availability of the CCC labor force were some of the reasons for the revitalization of water harvesting on rangelands.

Fifty-five percent of these stock ponds had to be renovated approximately 50 years later in 1984. Maintenance costs for the original stock ponds during the 50 year period before renovation was minimal due to limited erosion of structures and slow sedimentation rates behind the earthen dams.

3. Rationale for revisiting use on rangelands

Though an ancient practice, there are several reasons why water harvesting has been nearly abandoned as a management tool across most land areas of the southwest U.S. as well as other arid to semiarid regions of the world. These reasons include: 1) a perceived notion that installation of water harvesting infrastructure is too expensive in relation to the resultant benefits, 2) legislative restrictions and their associated costs for applications in the public land ownership landscapes of the western U.S., 3) a persistent belief that large spatial scale installations are too difficult to implement and maintain, or 4) a lack of knowledge about the effectiveness of water harvesting as a legitimate management practice. In reality, installation of simple systems are not that expensive, minimal maintenance is all that is needed to maintain function, and water harvesting can be effective if given enough time to be activated in regions of sparse, sporadic, and spatially widespread rainfall. Though there may be a lack of communication about the details of these structures, this could be rectified by more effective documentation and educational outreach programs. There is also the possibility that there are falsely held ideas that these methods are only suitable for areas of extreme poverty and with little access to more modern technologies. However, the water conserving nature of these water harvesting methods should dispel this idea. As water scarcity continues to pressure our increasing population, the importance of water conservation through water harvesting will be much more relevant (Giordano, 2009).

To some degree, arid land water managers may have overlooked the fact that there were installations of numerous rangeland water harvesting treatments in the western U. S. and other parts of the world in the 1930-1970s. Results from these applications are useful in improving the understanding of the advantages of employing water harvesting technologies. The desirability of these water harvesting techniques should increase in the future under conditions of climate change and increasing climate variability. New Mexico, because of its southerly location in the United States, has already experienced warmer temperatures (+1° C in winter and +2° C in summer) as a result of the ongoing climate change (Watkins, 2006). Diffenbaugh et al. (2008) also noted that the southwestern U.S. stands out as a regional hot spot for 21st Century climate change. It is possible that certain locations of the arid southwestern U. S. will not only experience warming temperatures, but also declining rainfall amounts, thus, greatly increasing the relevancy of water harvesting approaches. Although some areas of the West may receive reduced annual rainfall, they could also experience increases of convective rainfall events. As a result, the water harvesting approaches which are effective during intense rainfall can be used to offset the effects of warmer temperatures and increased evaporative losses that would be expected.

4. Methods of rangeland water harvesting

The basic goal of water harvesting on rangeland is to intercept the flow of surface water, either as overland flow or as channel flow. A variety of surface structures have been used in the past, but use of earthen dikes, berms or dams has been most popular because of simplicity, effectiveness, and relative low cost of both installation and subsequent maintenance. The concept of water ponding is to use a dike or berm to hold the water in such a manner so that it cannot flow off the surface of the soil unless the capacity of the dike is exceeded (Miller et al., 1969). When a pond forms behind the dike, the infiltration process has an extended time period to operate and replenish the soil moisture reservoir. Furthermore, by slowing down the flow of water, the amounts of infiltration and soil moisture are increased. Because of the increase in soil moisture, plant growth can be enhanced, either from existing plants, germination and resulting establishment, native seed banks, or planting of seeds during construction of the dikes. Water harvesting methods to supply livestock drinking water employ the same general techniques used in water ponding dikes. In this case, the collected water is stored in tanks or ponds (Frasier, 2003).

Although not serving exactly the same function as water ponding dikes, earthen berms are also installed across large areas upslope of downstream areas that are prone to flooding. The purpose of these berms is to slow down surface runoff, promote infiltration, control erosion, reduce flash flooding peaks, and even out the flows reaching the stream channel so that adverse impacts on downstream reservoirs are minimized (Caird & McCorkle, 1946; Baquera, 2010).

Water spreaders used on rangeland usually cover larger areas than water ponding dikes and are generally of two kinds. The first is designed as a system of dikes or berms constructed to automatically divert storm flows in gullies and spread them over the adjacent rangeland to promote the growth of forage (Miller et al., 1969). Such water spreading systems can also be used effectively with irrigated agriculture. The second type of spreader is more specific and requires a water storage reservoir that retains water during storm runoff events. When a certain volume of water has been stored, the entire stored volume is released in a quick burst to run down a restricted flow path like a modified arroyo system. Earthen berms are used to cause the discharged water to flow through a more sinuous channel, longer than the natural arroyo channel. The resulting larger volume of water has a greater length to follow while infiltrating into the channel bottom of the target area. This also promotes increased soil moisture which can enhance plant growth.

Useful forage plants can be seeded along the flow path to produce an increase in forage for livestock and wildlife.

To increase the water volume available for release, flow in stream channels of adjacent watersheds can be diverted to the storage reservoir to more rapidly increase the stored water volume. The soil berms are sometimes reinforced with concrete, especially at bends in the sinuous channel, to prevent bank erosion due to the transport of high flows over a short period of time.

Although the concept of shallow water ponding dikes to enhance soil moisture and, subsequently, increase ground cover and forage for livestock and wildlife is simple in concept, many factors enter into their exact placement in arid and semiarid regions. The overriding purpose is to slow down surface runoff, and one consideration is to determine areas with significant overland flow. This can be done by observing such flows in the field during or after heavy rainfall events, but this requires on-the-ground observations during what may be rare runoff events. It may be more useful to employ remote sensing data either by observing the evidence of overland flow immediately after a rainfall event, or by recognition of overland flow paths during post-rainfall dry periods that remain highlighted for several weeks because of remnant surface soil moisture patterns.

Figure 6 is an aerial photograph at the Jornada in southcentral New Mexico in October 2006 which shows runoff flow paths through the desert (after rainfall events) as darker areas where surface soil moisture is greater. From a landscape perspective, the use of remote sensing allows a more complete understanding of the landscape units generating surface runoff. This more detailed spatial analysis improves the actual placement of individual dikes.

Fig. 6. Aerial photograph over the Jornada in October 2006 showing runoff flow paths after rainfall events at 25 cm resolution. Flow paths are darker because of an increase of surface soil moisture following overland flow.

The type of soils where dikes are constructed needs to be considered because of the differential amounts of overland flow that can be generated by different soils. Fine to medium texture soils generally produce significant surface runoff from intense rainfall that can be intercepted by water ponding dikes (Miller et al., 1969). Sandy soils allow higher rates of infiltration, generate too little surface flow, and are, therefore, unsuitable for producing enough water for installation of dikes. Clay, silty-clay, or silty loam soils are generally suitable soil types for water ponding dikes. Once the pond is formed behind a dike, it is important to have the water infiltrate into the soil and be stored in the soil moisture reservoir for plant uses. Water harvesting dikes also promote sediment deposition in the ponding area. Generally, this can result in increases of the clay, silt, and/or loam contents of the soil which may allow more stored soil moisture and greater vegetation production.

Mean annual and seasonal rainfall and the type and intensity of storms are important rainfall characteristics for designing any water harvesting system. According to Bennett (1939), if the mean annual rainfall is from 200-355 mm, the conditions are ideal for plant growth for rangelands using water ponding. If a large portion of the rainfall occurs in convective rainfall events in summer, the chance of successful water ponding increases because rainfall rates are more likely to exceed infiltration rates and produce more runoff than areas with many, low intensity storms. If the mean annual rainfall exceeds 355 mm, then water harvesting for supplemental feed and cultivated crops also has a high probability of success (Bennett, 1939). These are characteristics present in the southwest U.S. as well as in the vast arid to semiarid regions of the world.

When completed, the actual water ponding dike should have a round rather than a V-shaped top because the rounded crest is less affected by large animal impacts, such as livestock trampling. Broad-based dikes with a bottom width of 2-3 m are more stable than narrow dikes. Dike lengths at the Jornada ranged from 50-150 m (Rango et al., 2006).

Periodic maintenance to repair breaches in dikes is recommended (Stokes et al., 1954). For a variety of non-technical reasons, the Jornada dikes were not originally thought to be effective, and the dikes have not received any maintenance since being constructed in 1975. Although this is not the optimum situation, the dikes are still performing their water ponding and increasing vegetation functions despite the development of breaches through the earthen dikes.

The soil type where a stock pond is to be constructed should possess silt and clay components, if possible, because of the capability of these soil textures to be compacted to increase the stock pond bottoms resistance to infiltration. However, construction on sandy soils is even possible. The USDA Forest Service (1939) suggested a way to make stock pond bottoms more impervious without using fine grain soils or other construction techniques. Salt was placed on the dry bottoms of the stock ponds to attract cattle. In a short time, the cattle would trample the soil of the future pond bottom into a hard compact state, and the stock pond would be nearly water tight when filled.

At the Jornada, when the first of the CCC stock tanks was built, the construction was accomplished using both five teams of three horses each with one Fresno plow (see Figure 4) and one of the first motorized caterpillars (see Figure 5). Today earthen dams used to form stock ponds are constructed around the world using anything from modern earth moving equipment to a large number (133) of men and women working with handheld tools (Botts, 2009).

5. Discussion

5.1 Water ponding dikes

Figure 7 shows a temporal sequence of vegetation growth behind the a) Ace Tank dikes and b) the Taylor Well dikes at the Jornada. When records were being kept of ponding events behind the dikes (1978-1981), the Ace Tank dikes averaged 12 ponding events per year, whereas the Taylor Well dikes averaged 11 per year. In the case of both sets of dikes, the response of vegetation behind the dikes (shown in darker brown or red tones) to ponded runoff water was not immediate, taking about 10-12 years to react to sporadic precipitation events typical of the southwest U.S. These delayed responses are to be expected in dry regions whereas in more humid regions, the response times may be 1-2 years. A similar delayed response was detected by Peterson & Branson (1962) on water harvesting structures installed by the CCC between 1934 and 1942 in southwestern New Mexico and southeastern Arizona. Initial surveys of vegetation growth showed little response, but subsequent surveys 12 years later revealed that vegetation growth was substantially improved. In the arid Southwest, it will take longer for the characteristic rainfall events (high intensity but widely distributed storms) to occur in the vicinity of the collecting area for water ponding dikes and still longer for extensive vegetation growth to occur.

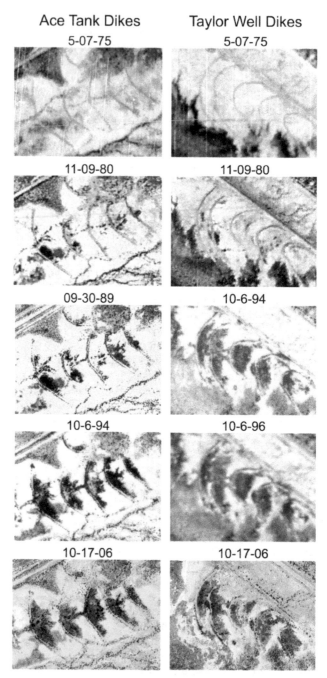

Fig. 7. A sequence of vegetation growth from construction in 1975 to 2006 using aerial photography for a) the Ace Tank dikes and b) the Taylor Well dikes at the Jornada.

Soil moisture was measured from the time of installation of the Jornada dikes in 1975 until when measurements were terminated in the mid 1980s. Tromble (1982) compared the soil moisture profile in July 1979 (after the dikes had been in place for four years) for the Taylor Well dikes (7.5 cm) and the Ace Tank dikes (15 cm). Although rainfall totals are similar for the Ace and Taylor dikes, the greater water ponding depth of Ace (due to the higher dikes) has produced a soil moisture profile difference. Figure 8 shows that the control area was uniformly dry down to 180 cm depth whereas the Taylor dikes were much wetter at the surface and gradually dried out with depth. The Ace dikes had uniformly greater soil moisture down to 180 cm (Tromble, 1982).

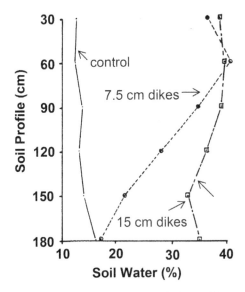

Fig. 8. Soil water profiles for control, 7.5-cm dikes, and 15-cm dikes on the Jornada Experimental Range on July, 1979 (from Tromble, 1982)

Associated with increases in soil moisture, Miller et al., (1969) have reported increased forage yields exceeding 1 ton/acre (2240 kg/ha) in response to water spreading treatments. Yields were reduced if water ponded without infiltrating for long periods of time. Branson (1956) reported that forage yields on water ponding dikes (as part of a water spreader system) were 2.6 times the yields on controls in a Montana experiment. Houston (1960), also working in Montana, reported an increase in herbage yields of 62% for water draining across rangelands, and a yield increase of 189% for rangeland where water was allowed to pond and infiltrate. Hubbell & Gardner (1950), experimenting in New Mexico, reported herbage yields increased by water spreading by 4-9 times and Hubbard & Smoliak (1953) reported herbage increases exceeding these yields. In the more recent water ponding experiments at the Jornada, Tromble (1984) reported that the 7.5 cm dikes resulted in a 2.4 – 6.0 fold increase in forage production over controls, depending on year and the location behind the dike. In all the water ponding or spreading experiments, it seems that increases in soil moisture and forage yield are consistent across the western U.S.

Few investigators have documented quantitative results and the costs and benefits associated with water harvesting for rangeland. Where this has been done, the investigators are usually more specific about costs but less so about the benefits. Investigators generally state that the costs are out-weighed by the benefits which are usually an increased amount of forage or increased plant species diversity and subsequent ground cover. Table 2 lists authors who have included costs (in U.S. $, for 2010) associated with installation of dikes or spreaders. Generally, the cost of construction of water ponding dikes is less than construction of water spreaders. The dikes used in water spreaders by Hubbard and Smoliak (1953) and Monson & Quesenberry (1958) range from 1.5-2.5 ft. (46-76 cm) high, whereas dikes employed at the Jornada ranged from 3-12 in. (7.5-30 cm) high. In order to estimate the costs of the Jornada dike installation (because no records were kept during installation), the dikes would be expected to cost about 50% of the average cost in Table 2 [0.5 x $32.48 = $16.24/acre ($40.10/ha)] in U.S. dollars) because the Jornada [average dike height = 7 in (18cm)] required much less construction effort than the ones reported in Table 2. To estimate the cost of installing dikes over a much larger area, eleven pastures at the Jornada have been identified as potentially feasible for trapping and ponding surface runoff because of favorable soil type and evidence of overland flow after convective storms. About 20% of the area in those pastures would be treated or about 3,641 acres (1,473 ha). The cost for water ponding dike installation would be approximately $59,000. Unfortunately, as with other analyses of the benefits and practices which can enhance goods and services from rangelands, we lack sufficient economic data for further cost/benefit calculations of the potential water ponding treatments.

Investigators	Cost / acre (1 ac = 0.405 ha)		Benefit
	Cost When Installed	Converted to 2010 $	
Mooney & Martin (1956)	$6.70	$53.13	% Increase hay
Hubbard & Smoliak (1953)	$0.36	$2.90	% Increase herbage
Monson & Quesenberry (1958)	$1.38	$10.29	% Increase herbage
Houston (1960)	$2.35	$17.10	Monetary increase of herbage/acre
Branson (1956)	$9.96	$78.99	Increased herbage/acre
Average		$32.48	

Table 2. Cost associated with installation of water ponding dikes or water spreaders.

5.2 Stock ponds

The number of stock ponds to be constructed depends upon the type of livestock to be grazed because of the distance that specific breeds of cattle typically graze away from a water source varies. For example, comparison studies using Global Positioning System (GPS) collars have shown that Mexican types and breeds of livestock such as the Criollo are willing to travel longer distances (5.8 km vs 2.2 km) from water than British breeds (Angus) when forage availability is limited. The Criollo will also travel to higher elevations than the Angus to seek out forage (Peinetti et al, 2011). Results from these studies indicate that the lighter, smaller Criollo breeds (~400 kg/animal) would probably be matched better than the

Angus breeds (~700 kg/animal) to the characteristics of desert rangelands (Peinetti et al, 2011). Designing stock water infrastructure for livestock breeds better suited to arid environments will result in reduced operational costs and reduced environmental impacts. For example, earlier studies by Herbel & Nelson (1966) found that Santa Gertudis cattle with a heritage from a hot, arid environment traveled 4.7 km a day further than Hereford breeds in search of forage. These cattle would seem to similarly disperse across the landscape like the Mexican breeds and reduce severe impacts around water sources as well as in areas with abundant forage.

When most of the stock ponds were constructed back in the 1930s at Jornada, a combination of manpower, horsepower, and mechanized vehicles were used (see Figures 4 and 5). The most usual stock tank is an earthen dam in a crescent shape that is constructed across an established drainage way. When original construction was done in 1934, the average time required for construction (with a crew of seven men, five Fresno teams of three horses each and one plow team) was two working days of eight hours each (USDA Forest Service, 1939). At that time the average cost of construction was $90 per stock tank ($1,450 in 2010). This value is considered to be on the low end of the range of construction costs because of the subsidized nature of using CCC labor. Construction costs in the same area of New Mexico before CCC labor was available averaged $157 ($2,529 in 2010) (Talbot, 1926). This was for a dirt tank that would store an average capacity of 220 m^3 of water. It is assumed that the same basic stock tank would be constructed today but with more modern earth moving equipment. The cost of the equipment would be higher, but the time for construction would be reduced. In 1935, the cost of drilling a deep well needed to yield this amount of water would be $2,833 ($44,524 in 2010 or more because the water table has continued to drop since the 1930s resulting in the need for deeper wells).

6. Conclusions

Water harvesting is a methodology that has been used for over 9 millennia to concentrate, collect, and distribute water that normally would be inaccessible for applications in irrigated agriculture, individual domestic water supply, and rangeland management. Although used widely for agriculture and domestic supplies, water harvesting is a management technique seldom used for rangeland applications despite numerous positive results. It is possible that the technique may be overlooked for a variety of possible reasons, the least compelling being that it is an ancient method based on archaic technologies.

As more and more stresses are placed on our natural resources through effects of a growing population, increased pressure on existing groundwater supplies, and climate change, a renewed use of water harvesting would have positive outcomes. The simplest technique is to use water ponding dikes which slow down surface runoff, allow infiltration and increase soil moisture, and promote significant vegetation growth for habitat cover and forage. It is recommended to use water ponding dikes because of the direct response: shallow water ponds form after high intensity rainstorms, infiltration and soil moisture increase, and growth of native vegetation (sometimes delayed for years because of the type and distribution of rainfall experienced across an area) is enhanced. The advantages of water ponding dikes are that they are simple to install, cost effective, and make use of water that would be lost to evaporation.

The use of water ponding dikes also mimics nature in the way that banded vegetation is arranged on the landscape: bare soil producing surface runoff after a storm, vegetation bands downslope slowing down and catching the surface runoff and increasing soil moisture, and causing increased vegetation growth as if it was located in an area with a higher rainfall. Future experiments are needed on larger areas to determine if these rangeland treatments cause improved vegetation cover that can expand to (or at least be stable over) even larger spatial extents. If water can be supplied effectively to the soil and vegetation complex, such as through water harvesting, it is likely that rangeland restoration projects will have an increased likelihood of success.

The use of stock tanks as water sources on rangeland for cattle grazing is a traditional method that has one of the least expensive construction costs amongst a variety of possible methods. In arid regions, it places reliance on trapping surface runoff that will otherwise be lost back into the atmosphere through evaporation. By confining this surface runoff in a pond with small surface area, and water depth of up to 2m, evaporation and infiltration losses are both reduced over what would normally occur if the water was spread out and infiltrated into a stream bed. The groundwater reservoir is not depleted until necessary, e.g., when severe drought years are encountered, and then well pumping would only be relied upon when the pond becomes dry.

7. Acknowledgements

The authors would like to thank scientists and technicians at the Jornada who have participated in rangeland remediation and water harvesting treatments. In particular, the late Dr. Robert Gibbens was an important source of information about the water harvesting research. His knowledge was invaluable for preserving a record of this type for future rangeland researchers. The technical assistance of Bernice Gamboa and Valerie LaPlante in preparing this manuscript is greatly appreciated.

8. References

Baquera, N. (2010) Characterization of Historical Water Retention Structures and Assessment of Treatment Effects in the Chihuahuan Desert. Master of Science Thesis, Department of Plant and Environmental Science, New Mexico State University, Las Cruces, New Mexico, USA, 69 pp.

Bennett, H. (1939). *Soil Conservation*. McGraw-Hill Co., New York, New York. 993 pp.

Boers, T. & Ben-Asher, J. (1982). A Review of Rainwater Water Harvesting. *Agricultural Water Management*, Vol. 5 pp. 145-158.

Botts, F., (2009). In: Excerpts from Journeys for a Witness by Florita Botts/ An earth Dam – Handmade in Ethiopia. http://www.drylandfarming.org/FB/FloritaHome.html. Accessed 6-28-2010.

Branson, F. (1956). Range Forage Production Changes on a Water Spreader in Southeastern Montana. *Journal of Range Management*, Vol. 9, pp. 187-191.

Branson, F.; Gifford, G.; Renard, K. & Hadley, R. (1981). *Rangeland Hydrology*, Range Science Series, Kendall/Hunt Publishing Company, Dubuque, Iowa, 339 pp.

Bruins, H.; Evenari, M. & Nessler, U. (1986). Rainwater Harvesting Agriculture for Food Production in Arid Zones: The Challenge of the African Famine. *Applied Geography*, Vol. 6, pp. 13-32.

Caird, R. & McCorkle, J. (1946). Contour-Furrow Studies Near Amarillo, Texas. *Texas Journal of Forestry*, Vol. 44, pp.587-592.

Caran, S. & Neely, J.(2006). Hydraulic Engineering in Prehistoric Mexico. *Scientific American*, Vol. 295, pp. 78-85.

Critchley, W., K. Siegert, C. Chapman, and M. Finkel. 1991. *A Manual for the Design and Construction of Water Harvesting Schemes for Plant Production*, AGL/MISC/17/91, Food and Agriculture Organization of the United Nations, Rome.

Cunningham, G.; Quilty, J. & Thompson, D. 1974. Productivity of Water Ponded Scalds. *Journal of Soil Conservation*, Soil Conservation Service of N.S.W, Vol. 30, Issue 4, pp.185-200.

Diffenbaugh, N.; Giorgi, F. & Pal, J. (2008). Climate Change Hotspots in the United States. *Geophysical Research Letters*, Vol. 35, L16709, doi:10.1029/2008GL035075.

Evenari, M.; Shanan, L. & Tadmor, N. (1982). *The Negev: The Challenge of a Desert*. Harvard University Press, Cambridge, Massachusetts, USA, Second Edition, 437 pp.

Fidelibus, M. & Bainbridge, D. (1995). *Microcatchment Water Harvesting for Desert Revegetation*. SERG Restoration Bulletin 5, SERG Soil Ecology and Restoration Group, San Diego State University, San Diego, CA, 12 pp.

Frasier, G. (2003). Livestock, Water Harvesting Methods for Encyclopedia of Water Science, Marcel Dekker, New York, pp. 593-595.

Frasier, G. & Myers, L. (1983). *Handbook of Water Harvesting*, Agricultural Handbook Number 600, Agricultural Research Service, U.S. Department of Agriculture, Washington, D.C., USA, 46 pp.

French, N. & Hussain, J. (1964). *Water Spreading Manual, Range Management Re. 1*, Pakistan Range Improvement Scheme, Lahore, Pakistan.

Giordano, M. (2009). Global Groundwater Issues and Solutions. *Annual Review of Environment and Resources* ,Vol. 34, pp. 153-178.

Hadley, R.; McQueen, I.; & et al. (1961). *Hydrologic Effects of Water Spreading in Box Creek Basin Wyoming*. U.S. Geological Survey, Water Supply Paper 1532A, U.S. Government Printing Office, Washington, D.C. 47 pp.

Havstad, K.; Huenneke, L. & Schlesinger, W. (2006). *Structure and Function of a Chihuahuan Desert Ecosystem: the Jornada Basin Long-Term Ecological Research Site*, Oxford University Press, New York, New York, 465 pp.

Heady, H. & Child, R. (1994). *Rangeland Ecology and Management*, Westview Press, Boulder Colorado, 519 pp.

Herbel, C. & Nelson, A. (1986). Activities of Hereford and Santa Gertrudis Cattle on a Southern New Mexico Range, *Journal of Range Management*, Vol. 19, pp. 173-176.

Holechek, J.; Pieper, R. & Herbel, C. (1995) *Range Management: Principles and Practices*, Prentice Hall, Englewood Cliffs, New Jersey, 526 pp.

Houston, W. (1960). Effects of Water Spreading on Range Vegetation in Eastern Montana. *Journal of Range Management*, Vol 13, pp. 289-293.

Hubbard, W. & Smoliak, S. (1953). Effects of Contour Dykes and Furrows on Short-Grass Prairie. *Journal of Range Management* Vol. 6, pp.55-62.

Hubbell, W. & Gardner, J. (1950). *Effects of Diverging Sediment Laden Runoff from Arroyos to Range and Crop Lands*. U.S. Department of Agriculture. Tech. Bull. 1012. 83 pp.

Hudson, N.W. (1987). *Soil and Water Conservation in Semi-Arid Areas*. FAO Soils Bulletin 57, Food and Agriculture Organization of the United Nations, Rome.

Lowdermilk, W. (1960). The Reclamation of a Man-Made Desert. *Scientific American* Vol. 202, pp. 55-63.

Miller, R.; McQueen, I.; Branson, F.; Shown, L. & Buller, W. (1969). An Evaluation of Range Floodwater Spreaders. *Journal of Range Management* Vol. 22, pp. 246-247.

Monson, O. & Quesenberry, J. (1958). *Putting Flood Waters to Work on Rangelands*. Montana Agricultural Experiment Station Bull. 543, 39 pp.

Mooney, F. & Martin, J. (1956). Water Spreading Pays-A Case History from South Dakota. *Journal of Range Management* Vol. 9, pp. 276-278.

Peinetti, H.; Fredrickson, E.; Peters, D.; Cibils, A.; Roacho-Estrada, J. & Laliberte, A. (2011). Foraging behavior of heritage versus recently introduced herbivores on desert landscapes of the American Southwest. *Ecosphere* Vol. 5, Issue 2, Article 57. (doi:10.1890/ES11-00021.1)

Peterson, H. & Branson, F. (1962). Effects of Land Treatments on Erosion and Vegetation on Rangelands in Part of Arizona and New Mexico. *Journal of Range Management*, Vol. 15, pp. 220-226.

Prinz, D., and A.H. Malik. 2002. *Runoff Farming*. WCA InfoNet, Rome, Italy, 39 pp.

Rango, A.; Tartowski, S.; Laliberte, A.; Wainwright, J. & Parsons, A. (2006). Islands of Hydrologically Enhanced Biotic Productivity in Natural and Managed Arid Ecosystems, *Journal of Arid Environment*, Vol. 65, pp. 235-252.

Renner, H. & G. Frasier, G. (1995). Microcatchment Water Harvesting for Agricutlural Production: Part I: Physical and Technical Considerations. *Rangelands*, Vol. 17(3), pp. 72-78.

Stokes, C.; Larson, F. & Pearse, C. (1954). *Range Improvement Through Waterspreading*. U.S. Government Printing Office. Washington, D.C. 36 pp.

Talbot, M. (1926). *Range watering places in the Southwest*. Department Bulletin, U.S. Department of Agriculture, Washington, D.C., 43 pp.

Tongway, D.; Valentin, C. & Seghieri, J. (2001). *Banded Vegetation Patterning in Arid and Semiarid Environments*. Springer-Verlag, New York, New York. 251 pp.

Tromble, J. (1982). Water Ponding for Increasing Soil Water on Arid Rangelands. *Journal of Range Management*, Vol. 35, pp.601-603.

Tromble, J. (1983). Rangeland Ponding Dikes: Design Criteria. *Journal of Range Management*, Vol. 36, pp. 128-130.

Tromble, J. (1984). Concentration of Water on Rangeland for Forage Production. *Second Intermountain Meadow Symposium Proceedings*, pp. 143-147. Gunnison, Colorado, Special Series No. 34

USDA Forest Service. (1939). *Notes from the Jornada Experimental Range*. Jornada Water Developments, Las Cruces, New Mexico, 2 pp.

Valentine, K. (1947). *Effect of Water-Retaining and Water-Spreading Structures in Revegetating Semidesert Range Land*. New Mexico College of A. & M.A., Agricultural Experiment Station Bulletin 341, 22 pp.

Vallentine, J. (1989). *Rangeland Development and Improvements*. Academic Press, San Diego, California, 524 pp.

Warren, J. (1965). The Scalds of Western New South Wales – a Form of Water Erosion. *Australian Geographer*, Vol. 9, pp. 282-292.

Watkins, A. (ed) (2006). *The Impact of Climate Change on New Mexico's Water Supply and Ability to Manage Water Resources*, New Mexico Office of the State Engineer, Santa Fe, New Mexico, USA.

Zaunderer, J. & Hutchinson, C. (1988). *A Review of Water Harvesting Techniques of the Arid Southwestern U.S. and North Mexico*. Working paper for the World Bank's sub-Saharan water harvesting study.

3

Performance Assessment and Adoption Status of Family Drip Irrigation System in Tigray State, Northern Ethiopia

Nigussie Haregeweyn[1,2], Abraha Gebrekiros[2], Atsushi Tsunkeawa[1],
Mitsuru Tsubo[1], Derege Meshesha[1] and Eyasu Yazew[2]
[1]Arid Land Research Center, Tottori University, Hamasaka,
[2]Department of Land Resources Management and Environmental Protection,
Mekelle University, Ethiopia
[1]Japan
[2]Ethiopia

1. Introduction

Large irrigation systems are generally incompatible among most of the African smallholder farming systems (De Lange, 1998) for the reason that support services for farmers, such as extension and credit are ineffective also often alter the established patterns of land tenure and land settlement, and have the effect of disrupting or undermining the established economic institutions. Moreover, improvements of surface irrigation may not be enough and due to the limited volumes of water harvested and stored as compared to crop water requirements. As a result, drip irrigation is being considered as one of the alternatives in the planning of irrigation in Tigray Regional State. Drip irrigation is often promoted as a technology that can conserve water, increase crop production, and improve crop quality. To this end, efforts to improve irrigation efficiency through new technologies have been undertaken in many areas of the Tigray Regional State. As a result, collaboration was initiated to improve irrigation efficiency between water-sector development organizations in Tigray, Ethiopia, and India, by the Norwegian Development Fund (DF) through the Triangular Project (Kirsten et al., 2008). As part of this effort, a family drip irrigation technology has been transferred from Gujarat in India to Tigray in Ethiopia, and it has spread throughout the region. There is now even a factory in Tigray that is producing the family drip irrigation (FDI) kits required for the drip irrigation.

Consequently, both governmental and non-governmental organizations have been providing technical and financial support to the family FDI beneficiaries in recent years. In this regard, selected individual household heads received one motor or pressurized treadle pump, one water tanker made of tin material having a capacity of 400 litters and a set of family drip kits to develop a 500 m² area. In addition, agricultural inputs such as vegetable seeds, fruit seedlings and fertilizer are also provided. All inputs are supplied on credit basis, which is to be paid back over a period of five years. The crops being grown were selected based on market demand.

Therefore, knowledge about performance of new irrigation technology is essential, since it serves as a base for formulation of irrigation projects. Moreover, it gives a bench mark for monitoring progresses within a given irrigation system. Thus, there is a need to evaluate the current small-scale irrigation in general and family drip irrigation in particular in the Tigray Regional State. The family drip irrigation technology is a case in point that has not been assessed in light of its technical, operational performance and adoption status. Such an assessment could assist in identifying constraints for future strategies that address water scarcity and consequently food security issues at household and national levels. Therefore, this study was undertaken to assess the performance and identify the factors that influence the technical and operational performance of family drip irrigation technology as well as its adoption status.

2. Materials and methods

2.1 Description of the study area

This study had two levels of investigation with regard to addressing the specific objectives of the study. The first was on assessment of the technical and operational performance of FDI system which was conducted in Wukro District located at 45 Km north of Mekelle, Eastern Zone of the Tigray Regional State, northern Ethiopia, while the second was on assessment of dissemination and adoption of FDI system at Tigray state's level (Fig. 1).

Fig. 1. Location of the Tigray Regional State and the Wukro district, case study area, Northern Ethiopia

The Wukro District is located between 13°14′ and 14°02′ N and 39°34′ and 39°37′ E, and covers a total area of about 100,228 ha (BoARD, 2008) (Fig. 1). It has 63 Kushets (the smallest administrative unit in the area), a total population of 99,688 with 23,899 household heads (FDREPCC, 2008). Based on analysis of climate data obtained from the National Meteorological Services Agency, the mean annual rainfall at Wukro station for the years 1996-2006 was computed at 723 mm. While the mean annual average daily air temperature is 20°C, with mean annual daily minimum and maximum values of 11°C and 28°C, respectively. According to IFSP (1995), the elevation ranges between 1960 and 2600 m.a.s.l. The same report classified the most common soil types of the area into three: Clay (*Baekel*), Red (*Mekayih*) and Black (*Walka*) .

2.2 Field layout of the FDI system
The installation of family drip irrigation had the following details:
- Four lateral pipes of 25 m long each with drippers spaced at 0.3m and a 20 m length manifold with laterals fitted along it at a spacing of 0.75m interval were laid out to irrigate an area of 500 m². The FDI kit contains a water tank, a riser, a manifold, laterals, disc filter, valve, end-cap, and different fittings, where the detailed experimental setup is given in Figure 2;
- The water container was placed at 2m high on a stand made of wooden logs;
- The height of water outlet was at 2.1m since the barrel was drilled at 10 cm above the bottom and a 3/4″ socket was welded to it in order to provide the water pressure required to operate the system and to allow suspended materials to settle;
- A 3/4″ disc filter was attached to a 3/4″ straight adaptor on the barrel;
- On the straight adaptor a 3/4″ disc filter was attached. Again a female elbow was connected to a disc filter and to the risers followed by flow regulator up on the manifold at both left and right sides;
- The manifolds having a diameter of 25 mm were laid and connected to the riser at the center, while the riser was connected to the elbow then to the water container;
- The laterals having a diameter of 12 mm were unrolled and laid along the full length of each bed of which plants to be irrigated;
- The laterals were connected with the manifold with the help of connectors at 0.75 m interval;
- The end of the laterals and the manifold were closed with end caps and they were tied at the beginning as well as at the end on wooden pegs to avoid direct soil contact and thus to prevent clogging as a result of back siphoning.

2.3 Performance assessment of the FDI system
2.3.1 Uniformity
Uniformity in here referred to what extent the water was uniformly distributed across the irrigated area. This is affected by the water pressure distribution in the pipe network and by the hydraulic properties of the emitters used (Smajstrla et al., 2002).

To measure emitter flow rates, a standard graduated cylinder was used and volume of water was collected for one minute. A stopwatch was used to measure the time. For uniformity determination, measurements were made three times at 32 locations in each FDI kits during the crop growing season taking into consideration the minimum recommended number points by Smajstrla (2002). Care was also taken to distribute the measurement points throughout the irrigated zone. In order to address all these issues, uniformity evaluation procedures developed by FAO (1980) were followed with the following details and also as shown in Figure 3;

Fig. 2. Detailed experimental set up for the family drip irrigation system of this study

Four laterals were located along an operating manifold; one at the left end of the inlet, one at the right end of the inlet and two located at ½ ways to the right and left from the inlet midpoint;

On each lateral, 2 adjacent emitters were selected at 8 different emitter locations: at the inlet end, 1/3 down way of the lateral, 2/3 down way of the lateral and at farthest end point of the lateral.

Four commonly used uniformity parameters were determined for this study:

The first uniformity parameter used was emission uniformity (E_U) which is the most useful system performance indicator for trickle systems. Sometimes in case of field evaluation it is defined as distributions uniformity (D_U) which was calculated using Equation 1 (Kruse, 1978).

$$E_u = 100 \times \left(\frac{q_{min}}{q_a} \right) \tag{1}$$

Where:

E_U = Field emission uniformity (%);

q_{min} = Minimum discharge rate computed from minimum pressure in the system (l/h) and

q_a = average of all field data emitter discharge rates (l/h).

The second uniformity parameter used was emitter flow variation (q_{Var}) which was calculated using Equation 2 (Wu, 1983):

$$q_{var} = \left(\frac{q_{max} - q_{min}}{q_{max}} \right) \tag{2}$$

Where:

q_{max} = maximum emitter flow rate (l/h) and

q_{min} = minimum emitter flow rate (l/h).

The third uniformity parameter used was coefficient of variation (C_V) which was calculated using Equation 3 (Wu, 1983).

$$C_v = \frac{S}{q_a} \qquad (3)$$

Where:

S = standard deviation of emitter flow rates (l/h) and

q_a = average emitter flow rate (l/h).

The fourth uniformity parameter used was uniformity coefficients (UC) which is often described in terms of the coefficient of variation defined as the ratio of the standard deviation to the mean and was calculated using Equation 4 (ASAE, 1985) expressed as:

$$U_c = \left(1 - \frac{S_q}{q_a}\right) \times 100 \qquad (4)$$

Where:

U_C = uniformity coefficient (%);

S_q = average absolute deviation of all emitters flow from the average emitter flow (l/h) and

q_a = average emitter flow rate (l/h).

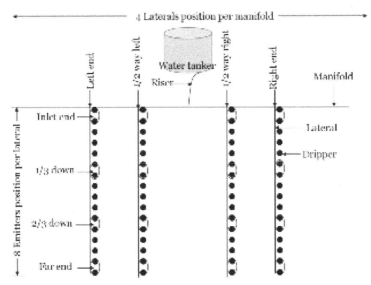

Total plot area 500 m². Spacing between laterals = 0.75m and drippers = 0.3m, manifold and lateral length 20m & 25m

Fig. 3. Layout of control points and discharge measurement for family drip irrigation system of this study

2.3.2 Determination of total suspended solids (TSS)

Total suspended solids (TSS) analytical test was employed to determine current or future potential emitters clogging problems arising from poor water quality. Water samples were taken from representative three different shallow wells after operating the motor pump, assumed as the worst case of water physical quality during water delivery moment in time. Taking into consideration the recommendation given by Clesceri et al. (1998) a 250 ml of water samples from each selected shallow wells were taken and oven dried at 105°c for 1 hour at the Soil Physics Laboratory Mekelle University.

The TSS was then calculated using Equation 5 and evaluated based on the Water Quality Guidelines developed by Hanson et al. (1994):

$$TSS = \frac{(A - B) \times 1000}{total volume} \tag{5}$$

Where:

A = weight of filter + dried residue (mg), and
B = weight of filter (mg).

2.3.3 Evaluation of the water supply-demand for FDI system for selected test crops

Assessment of the existing water supply and the crop water requirements of the two dominantly cultivated crops (onion and tomato) as test crops were done. The total amount of water supplied to each crop throughout the growing season was assessed by multiplying the amount of water applied per irrigation and the frequency of irrigation. The daily volume of water supplied by the farmer to the test crops were taken from farmers' current operation practice. The irrigation frequency was found to be two times per day; one in the morning and the other in evening with total daily supply volume of 0.4m³ water. Taking into account reference evapo-transpiration (ET₀), crop type, length of growth, growth stage and effective rainfall, gross irrigation requirement was computed for the two test crops. An average daily ET₀ 5.12 mm/day as determined by Haftay (2009) was used for this study. The crop water requirement for the two test crops was estimated by applying Equation (6) given as:

$$ET_c = (ET_o \times K_c) \tag{6}$$

Where:

ET_c = crop evapotranspiration;
ET_o = reference evapotranspiration and
K_c = crop coefficient values which were adapted from Doorenbos and Pruitt (1977).

The net irrigation requirement (NIR) was computed using Equation 7 given as:

$$NIR = (ET_c - P_e) \tag{7}$$

Where:

ET_c = crop evapotranspiration and
P_e = effective rainfall.

Gross irrigation requirement (GIR), which is defined as the depth or volume of irrigation water required over the whole cropped area excluding contributions from other sources, plus water losses and /or operational wastes was estimated using Equation 8 (FAO, 1980) as:

$$GIR = \left(\frac{NIR}{E_a} \right)$$ (8)

Where:

GIR = gross water requirement and

E_a = the application efficiency, assumed to be 90% as an attainable value of application efficiency for drip irrigation.

2.4 Assessment of FDI kits dissemination trend and adoption Status

To understand the adoption and dissemination status across the region, it was essential to know the spatial and temporal distribution of the system first . For this, a list of distributed family drip irrigation kits over the period of 2004-2008 was obtained from the Tigray Regional Bureau of Agriculture and Rural Development (BoARD), the Tigray Bureau of Water Resource Development (BoWRD) and the Relief Society of Tigray (REST), local development organizations operating in irrigation development in the region. Furthermore, the records obtained from the three Bureaus were organized based on spatial and temporal sequences. In addition, the delivered FDI kits were identified as installed and uninstalled to understand their working conditions.

While for analysis of FDI adoption status and rate, a three-stage sampling techniques were employed to collect data. Accordingly, random samples of 120 household heads were selected from three sites (*Tabias*). Each site consisted of 40 randomly selected respondent farmers from both users and non-users of FDI technology. Besides this, a two-part questionnaire was developed. The first questionnaire consisted of project structural evaluation based on attitudinal or knowledge statements about FDI technology, with possible responses and explanations by the respondent farmers. While the second questionnaire consisted of questions dealing with demographic, education level, age, and source of water and related characteristics of the respondents to identify and analyze variables that were supposed to influence FDI technology adoption. The content of the questionnaire was designed using inputs from staff members of the governmental and non-governmental organization, especially working with the FDI system technology including FDI user farmers. Rejection and inclusion of the variables was made based on the required expected frequency and related criteria as suggested by Rangaswamy (1995). Finally, the adoption status and rate were analyzed using a Chi-square test statistics of the contingency table at significance levels of $P < 0.05$ and 0.01.

3. Results and discussion

3.1 Performance assessment of the FDI system

3.1.1 Uniformity

The uniformity parameters (emission uniformity, flow variation, and uniformity coefficient) values of the three selected FDI systems are given in Table 1. The average E_U values for the selected FDI systems were 93.67%, 93.85% and 94.34% respectively (Table 1). The emission uniformity obtained from the experiment were found better as compared to the findings by Polak and Sivanappan (2004), for low-cost drip systems using holes made with a heated punch as emitters that reported uniformity rate of 85%. While systems using micro-tubes had uniformity rates of approximately 90%. According to ASAE (1985) standards and other experimental results of FAO (1984), on the general criteria for emission uniformity, emission uniformity greater than 90% is characterized as an excellent range of performance.

A flow variation (q_{var}) values of 6.8%, 6% and 5% were obtained for FDI$_1$, FDI$_2$ and FDI$_3$ respectively. According to Braltes (1986), general criteria for emitter flow variation gives as <= 10% desirable, 10-20% acceptable and >20% unacceptable ranges. Thus, this field-based result showed that the performances of all the three FDI system observations were within the desirable range of recommendation which were having less than 10% emitter flow variation. Moreover, a mean coefficient of variation (CV) for flow variation (q_{var}) values of 0.34, 0.27 and 0.17 were obtained for FDI$_1$, FDI$_2$ and FDI$_3$respectively. This indicated that the results obtained in this experiment were marginal to unacceptable for FDI$_1$ and average for FDI$_2$ and FDI$_3$ based on the guidelines set up by the American Society of Agricultural Engineers ASAE (1985).

Average uniformity coefficient (Uc) values of 73%, 97% and 98 % were obtained for FDI$_1$, FDI$_2$ and FDI$_3$ respectively. These values indicate that FDI$_2$ and FDI$_3$ systems were found to have a uniformity coefficient values rated as excellent (> 90%), but the uniformity coefficient value for FDI$_1$ was below 85%, which was considered as rationally bad range of performance as suggested by Malik et al. (1994).

In general the different aspects of the FDI uniformity indexes used in this study revealed that the FDI technology has no as such significant problem in relation to non-uniform water distribution within the field.

FDI$_1$	parameters			
observed	$E_{U\,(\%)}$	$q_{Var\,(\%)}$	$C_{V(ratio)}$	$UC_{(\%)}$
Beginning	94.04	5.00	0.21	99.79
Middle	93.91	8.50	0.54	46.00
End	93.06	7.00	0.26	74.00
FDI$_2$				
Beginning	94.05	3.00	0.02	98.00
Middle	95.14	10.00	0.02	98.00
End	92.35	5.00	0.04	96.00
FDI$_3$				
Beginning	95.17	3.00	0.01	99.00
Middle	94.74	7.00	0.02	98.00
End	93.12	5.00	0.02	98.00

Eu: Emission uniformity; q_{var}: Flow variation; C_V: Coefficient of variation; U_C: Uniformity coefficient.

Table 1. Uniformity parameter values of the three selected FDI systems

3.1.2 Total suspended solids (TSS) and emitter clogging hazards

Results of the TSS analytical test showed 144, 116 and 96 mg/l for shallow wells 1, 2 and 3 respectively (Table 2). According to Water Quality Guideline for micro irrigation developed by Haman et al. (1987), the TSS results in this study fall in a moderate to severe grounds for emitter clogging hazards. As shallow wells 1 and 2 are where a severe clogging problem is likely to occur it calls for pre-filtration or improve filtration mechanisms within the system before emitter plugging hazard occurs.

Pan No	Sample	Mass pan+	Volume of water sample	Mass pan+	mass TSS	TSS (mg/L)=
	code	filter (gm)	(ml)	filter + TSS (gm)	(gm) = [e-c]	[f/d] x 10^6
a	b	c	d	e	f	g
1	shallow well 1	2	250	2.036	0.036	144
2	shallow well 2	2	250	2.021	0.029	116
3	shallow well 3	2	250	2.024	0.024	96

Table 2. Total suspended solids (TSS) for the three shallow wells.

3.1.3 Evaluation of the water demand and supply for FDI system

The estimated total water requirements for onion and tomato were 315 m^3 and 180 m^3 while the corresponding total water supply was 120 m^3 and 96 m^3 respectively. Furthermore, the daily water demand for plot size of 500 m^2 is 2.1 m^3 for onion and 1.53 m^3 for tomato (Table 3).

From this result, the farmers need to apply the required quantity of water for the crop, and for that they need to be aware of the supply-demand relationships through organizing demonstrations and trainings. In case, labor availability is a problem to cover the entire area, they may reduce the size of the irrigated plot from 500m^2 to 190 m^2 for onion and 27 m^2 for tomato, respectively. Failure to supply the required amount of water to the crop would result in a significant yield reduction, which could eventually force the farmers to abandon the use of FDI system technology.

crop Type	D.W.R (mm/d)	G.W.R (m³/A)	T.W.R (m³/A)	Area (m²)	D.W.S (mm/d)	T.W.S (m³/A)	Deficit (m³/A)	Deficit (%)
Onion	4.14	2.1	315	500	0.8	120	-195	61.9
Tomato	3.06	1.53	183.6	500	0.8	96	-87.6	52.28

D.W.R: Daily water requirement; D.W.S: Daily water supply; T.W.S: Total water supply; G.W.R: Gross water requirement; d: Day ; A: Area.

Table 3. Comparison of water demand and supply for Onion and Tomato crops.

3.2 Assessment of FDI kits dissemination trend and adoption status
3.2.1 Distribution trends of FDI system kits

Figures 4 & 5 show that the distribution of FDI kits has shown increasing trend both across the years and zones. However, sites assessment results showed that, there was a variation in FDI kit supply within a given time and place in all Zones of the region. Analysis of the distribution records in the past 5 years (2004-2008) shows that, the maximum FDI kit distribution was observed in year 2008. The established factory that is producing the equipment required for drip irrigation system may have a significance contribution in maximizing the temporal and special distribution trends of the technology.

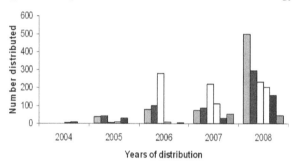

Fig. 4. Temporal distribution trend of FDI system at zonal Level of the Tigray Regional State, Northern Ethiopia

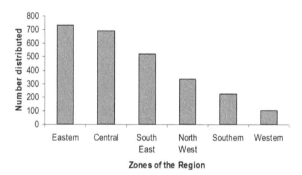

Fig. 5. Spatial distribution trend of FDI system at zonal level of the Tigray Regional State, Northern Ethiopia.

However, the number of working (installed) FDI Kits throughout region were only 1442 out of the 2615 supplied (i.e. 55 %). There is high spatial variation among the zones in the region which ranges between 20 % in Southern Zone to 84 %, in Southeast Zone (Figure 5). However, In Wukro district where this study was conducted, 100% the delivered FDI Kits were installed in the field (Figure 6). This shows that Southeast Zone relatively attained the satisfactory results in-terms of installing the delivered FDI kits at zonal level. Based on the findings, discussions and communications (formal and informal) held with beneficiaries, stakeholders, experts and administrators at different managerial levels during and between the assessments of FDI trends, those areas with low achievement of FDI installation were characterized by inadequate extension services, supervisions and monitoring the operational progress and low involvement of non-governmental organizations (NGOs). Since, the involvement of NGOs both in application of technique and operation of the delivered FDI kits might be their own contribution during the installation.

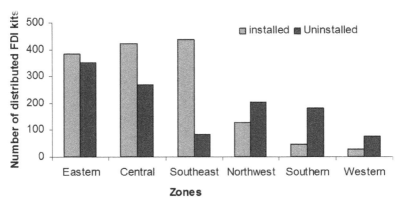

Fig. 6. FDI system distribution zones of the Tigray Regional State, Northern Ethiopia.

Conversely, the study area has no problem of installation for the delivered FDI kit. Though, extension services, monitoring and other related activities may have less importance, however, like other areas of the region, there is still variability in both temporal and spatial distribution of FDI system kits (Figures 7 & 8). Yet, there are two sites (Kihen and Debreberhan) among the 15 studied sites where FDI system intervention was absent.

In majority of the cases in the study area (District), sites (*Tabias*) with low to nil FDI system intervention were located outside of the main road of the District. These areas are also characterized by inadequate infrastructures such as access to roads, extension services, marketing outlets that attributed to the slow pace of FDI dissemination in the study area.

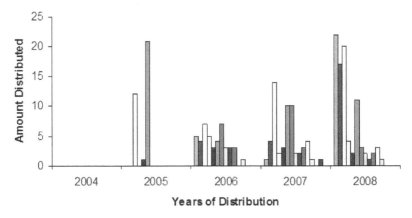

Fig. 7. Temporal Distribution Trend of FDI kit for 15 *Tabias* of Wukro Woreda in Tigray Regional Sate, Northern Ethiopia

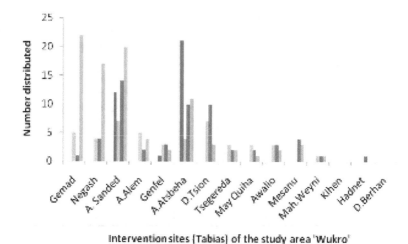

Intervention sites (Tabias) of the study area 'Wukro'

Fig. 8. Spatial Distribution Trend of FDI kit for 15 *Tabias* of Wukro Woreda in Tigray Regional Sate, Northern Ethiopia

3.2.2 Factors controlling adoption of FDI system

3.2.2.1 Age group and adoption status

Age group was found to influence the FDI adoption rate significantly ($P < 0.05$; Table 4). Younger farmers (30-45 years of age) were found relatively better adopters of FDI technology than older ones as the latter were not convinced with the significance of water drops to satisfy crop needs as compared to the one traditionally used furrow irrigation.

Age	FDI adoption status							
group	Current users		Current non-users		Future users		Total	
	No	(%)	No	(%)	No	(%)	No	(%)
30-45	20	54.1	9	24.3	8	21.6	37	100
46-60	10	37.1	8	29.6	9	33.3	27	100
60&above	8	14.3	40	71.4	8	14.3	56	100
Total	38	31.7	57	47.5	25	20.8	120	100

Table 4. Age group and FDI adoption status

3.2.2.2 Education level and FDI adoption status

Education level was found to influence adoption rate significantly ($P < 0.05$; Table 5). Farmers with exposure to primary school (grades 1-6) were found dominant adopters of FDI technology. Uneducated farmers were the lowest adopter. Therefore, in order to expand FDI technology utilization it would be sound to work with literate farmers in general and grade 1-6 in particular.

Education level	FDI adoption status							
	Current Users		Current non-users		Future users		Total	
	No	(%)	No	(%)	No	(%)	No	(%)
Non-educated	9	18.8	30	62.5	9	18.7	48	100
1- 6 grade	19	42.2	9	20.0	17	37.8	45	100
Grade 7 & above	10	37.0	10	37.0	7	26.0	27	100
Total	38	31.7	49	40.8	33	27.5	120	100

Table 5. Education level and FDI adoption status

3.2.2.3 Access to water source type and FDI adoption status

Farmers having access to shallow well water source were found better adaptors of FDI technology as compared to farmers having access to surface water source ($P < 0.01$; Table 6). This variability in adoption rate of the technology is related to the location of the water sources in relation to homesteads that made it easy to follow-up and manage the farm. Moreover, using shallow wells as source of water for FDI technology is relatively secured from vandalism of FDI kits because of the relative advantage being nearer to homesteads with that of surface water sources.

Water source	FDI users		FDI non-users		Total	
	No	(%)	No	(%)	No	(%)
Ground	31	83.8	6	16.2	37	100
Surface	7	8.4	76	91.6	83	100
Total	38	31.7	82	68.3	120	100

Table 6. Access to water source and FDI adoption status

3.2.2.4 Gender and FDI adoption status

Female-headed households were found better adopters of the FDI technology as compared to male-headed household heads though not significantly different (Table 7). The better adoption rate of female household heads may arise from their access to work around their homestead for long time. Moreover, the provision protocol of FDI kits encourages female household heads.

Sex	FDI users		FDI non users		Total	
	No	(%)	No	(%)	No	(%)
Female	10	40.0	15	60.0	25	100
Male	28	29.5	67	70.5	95	100
Total	38	31.7	82	86.3	120	100

Table 7. Gender and FDI adoption status

4. Conclusions

Household family drip irrigation technology has been introduced recently in the Tigray Regional State as an option to conserve water and hence to increase crop production in the region. This study evaluated its performance on the basis of various performance indicators.

Average uniformity coefficient values of 73 %, 97 % and 98 % were obtained for FDI_1, FDI_2 and FD_3 respectively. Based on ASAE (1985) criteria, the results obtained in this experiment were marginal to unacceptable for FDI_1, but good for FDI_2 and FDI_3. The clogging hazard was moderate to severe under current operation conditions of the FDI system, which may add up on the cost of spare parts and would likely to reduce the adoption rate by farmers. Therefore, regular inspection of emitters to identify clogged ones and undertaking of routine maintenances are necessary. Dismantling, blowing in it, or flashing out with water could help maintaining a clogged emitter. If, the situation is more serious, it is better to change the emitters. On-line type of emitter is more favorable than in-line ones because on-line emitters can be dismantled and repaired easily by the farmer. Frequent inspection and cleaning of filter is also more important.

Under the existing FDI operating condition, the supplies of water for the crops were very low to satisfy their demand. This indicates that, farmers and extension workers have limited knowledge and perception about the FDI technology operation systems. Thus, the users and development workers may need further training and demonstration of the technology at field level under farmers' operating condition. Moreover, appropriate technical and agronomic guidance and support to farmers in development and introduction of drip sets to sustain adopter's motivation throughout the season are needed.

The result of this field-based study revealed that the lower growth of FDI system utilization is not associated with the technology itself but it is rather due to the lack of awareness by the farmers and development agents on the technical and operational requirements of the FDI system to effectively operate and utilize the technology at household level.

Therefore extension services to raise awareness on the utilization and management, and mechanisms to monitor the development FDI technologies implementation should be strengthened. Moreover, further study is still needed to analyze the economic feasibility of the FDI system.

5. References

American Society of Agricultural Engineers 'ASAE'.1985. *Design, installation and performance of trickle irrigation systems.* ASAE standard EP 405, St. Joseph, Michigan, pp. 507-510.

Bureau of Agriculture and Rural Development 'BoARD' .2008. *A survey conducted in the annual report of the District (wukro)* office of Agriculture and Rural Development.

Barlts, V.F. 1986. *Operational principles-field performance and evaluation.* Trickle irrigation for crop production, Amsterdam, Elsevier, pp.216-240.

Clesceri L.S., Greenberg, A.E. Eaton, A.D. 1998. Method 2540D, *Standard Methods for the Examination of Water and Wastewater,* 20th Edition. American Public Health Association. Washington DC.

Doorenbos, J., Pruitt, W.O. 1977. *Crop Water Requirements.* FAO Irrigation and Drainage Paper, Bull. FAO n" 24, pp. 144.

De Lange M (1998). *Promotion of low cost and water saving technologies for small-scale irrigation. South Africa:* MBB Consulting Engineers.

FAO. 1980. *Localized Irrigation: Design, installation, operation and evaluation.* Irrigation and Drainage Paper, No. 36, FAO, Rome.

FAO. 1984. *Localized Irrigation: Design, installation, operation and evaluation.* Irrigation and Drainage Paper, No. 36, FAO, Rome.

FAO. 1998. *Institution and technical operations in the development and management of small- scale irrigation.* pp. 21-38. Proceedings of the third session of the multilateral cooperation workshops for Sustainable Agriculture, Forest and Fisheries Development, Tokyo Japan, 1995, FAO Water Paper, No. 17, Rome.

Federal Democratic Republic of Ethiopia Population Censes Commission F.D.R.E.P.C.C. 2008. *Population and housing census summary and statistical report of 2007.* Pp 54.

Haman, D.Z., Smajstrla, A.G., Zazueta F.S. 1987. *Water Quality Problems Affecting Micro irrigation in Florida.* Agricultural Engineering Extension Report 87-2. IFAS, University of Florida

Hanson, B.A., Fauton, D.W., May. D. 1995. *Drip irrigation of row crops*: An overview. Irrigation Science l, 45(3), Pp 8-11.

Haftay Abrha. 2009. *Crop water fertilizer interaction and physico-chemical properties of the irrigated soil.* Post graduate studies (unpublished). Mekelle University, Mekelle, Ethiopia.

Isaya, V.S. 2001. *Drip Irrigation: Options for smallholder farmers in Eastern and Southern Africa.* Regional Land Management unit (RELMA/SIDA), technical and book series 24, Nairobi, Kenya.

Integrated Food Security Program 'IFSP'. 2005. A study conducted in the five year development plan of the drought-prone areas of Tigray regional state districts. Mekelle, Tigray, Ethiopia.

Kruse, E.G. 1978. Describing irrigation efficiency and uniformity. J. Irrig. and Drain Div., ASCE 104 (IR1), pp. 35-41.

Kirsten, U., Sygna, L., O'brien K., .2008. *Identifying sustainable path ways for climate adoption and poverty reduction.* Pp - 44.

Keller, J., Keller, A.A. 2003. *Affordable drip irrigation systems for small farms in developing countries.* Proceedings of the irrigation Association Annual Meeting in San Diego CA, 18-20 November 2003. Falls Church, Virginia, Irrigation Association.

Malik, R.S., Kumar, K., Bandore, A.R. 1994. *Effects of drip irrigation levels on yield and water use efficiency of pea.* Journal of Indian Society Soil Science. Vol. 44, No. 3. Pp 508-509.

Polak, P., Sivanappan, R.K., 2004. *The potential contribution of low-cost drip irrigation to the improvement of irrigation productivity in India.* Indian water resources management sector review, report on the irrigation sector. The World Bank in cooperation with the Ministry of Water Resources, Government of India, pp 121-123.

Rangaswamy, R. 1995. Agricultural statistics, new age international publishers. Pp105-110

Smajstrla, A.G., Boman, B.J., Pitts, D.J. Zueta, F.S., 2002. *Field evaluation of micro irrigation water application uniformity.* Fla. Coop. Ext. Ser. Bul.265. Univ. of Fla.

Wu, I.P., 1983. A unit-plot for drip irrigation lateral and sub-main design. ASAE paper, St. Joseph, MI 49085. No. 83-1595.

4

Importance of Percolation Tanks for Water Conservation for Sustainable Development of Ground Water in Hard-Rock Aquifers in India

Shrikant Daji Limaye

UNESCO-IUGS-IGCP Project 523"GROWNET"
International Association of Hydrogeologists (IAH)
[1]Ground Water Institute
[2]Association of Geoscientists for International Development (AGID)
[3]International River Foundation, Brisbane,
[1]India
[2]UK
[3]Australia

1. Introduction

Development of a natural resource like ground water is a concerted activity towards its sustainable use for human benefit. The concept of sustainable use is related to various factors like the volume of water storage in the aquifer, annual recharge or replenishment, volume of annual pumpage for the proposed use, benefit/cost ratio of the proposed use, and environmental impacts of the proposed use.

Hard rock aquifers in this paper mean the non-carbonate, fractured rocks like the crystalline basement complex and metamorphic rocks, which cover an area of about 800,000 sq. Kms. in central and southern India. Basalts of western India also known as the Deccan traps of late Cretaceous to early Eocene period are also included as a special case. Deccan traps comprise hundreds of nearly horizontal, basaltic lava flows in a thick pile and cover around 500,000 sq. kms of western India. (Fig. 1a and 1b) This pile was not tectonically disturbed after consolidation and a hand specimen does not show any primary porosity due to the non-frothy nature of the lava. (Adyalkar & Mani 1971) Hydrogeologically, the Deccan traps have low porosity and are therefore, akin to fractured hard rock aquifers.

The most significant features of the hard rock aquifers are as follows:

1. A topographical basin or a sub-basin generally coincides with ground water basin. Thus, the flow of ground water across a prominent surface water divide is very rarely observed. In a basin, the ground water resources tend to concentrate towards the central portion, closer to the main stream and its tributaries.

2. The depth of ground water occurrence, in useful quantities, is usually limited to a hundred meters or so.

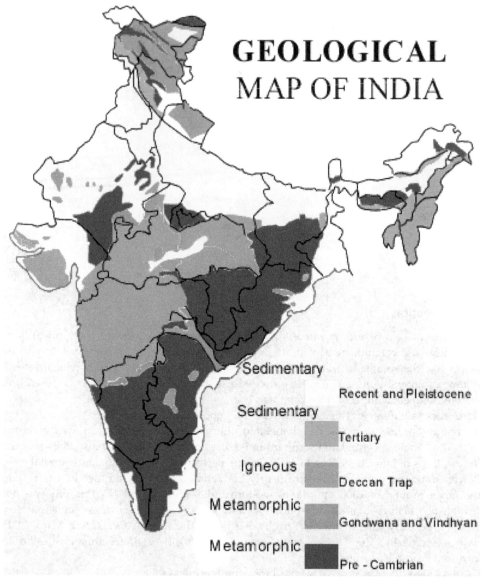

Note: Basement Complex (Granite, Gneiss, and other Metamorphic rocks) is shown in brown color. Basaltic area is shown in green. Black lines show State boundaries. The state of Maharashtra on the western coast is almost covered by Basalt or Deccan Trap (Green color). Peninsular India is mostly covered by Basalt and Basement Complex.

Fig. 1(a). Geological Map of India showing Basement Complex and Basalts of Peninsular Region. (Scale: 1cm = 200 km).

Fig. 1(b). Nearly horizontal lava flows comprising the Deccan Traps or Basalts of Western India. Location: Western Ghat hills between Pune and Mumbai, Maharashtra State.

3.   3. The aquifer parameters like Storativity (S) and Transmissivity (T) often show erratic variations within small distances. The annual fluctuation in the value of T is considerable due to the change in saturated thickness of the aquifer from wet season to dry season. When different formulae are applied to pump-test data from one well, a wide range of values of S and T is obtained. The applicability of mathematical modeling is limited to only a few simpler cases within a watershed. But such cases do not represent conditions over the whole watershed.

4.   4. The phreatic aquifer comprising the saturated portion of the mantle of weathered rock or alluvium or laterite, overlying the hard fractured rock, often makes a significant contribution to the yield obtained from a dug well or bore well.

5.   5. Only a modest quantity of ground water, in the range of one cu. meter up to a hundred cu. meters or so per day, is available at one spot. Drawdown in a pumping dug well or bore well is often almost equal to the total saturated thickness of the aquifer.

Ground water development in hard rock aquifer areas in India and many other countries has traditionally played a secondary role compared to that in the areas having high-yielding unconsolidated or semi-consolidated sediments and carbonate rocks. This has been due to the relatively poor ground water resources in hard rocks, low specific capacity of wells, erratic variations and discontinuities in the aquifer properties and the difficulties in exploration and quantitative assessment of the resource.

It should, however, be realized that millions of farmers in developing countries have their small farms in fractured basement or basaltic terrain. Whatever small supply available from these poor aquifers is the only hope for these farmers for upgrading their standard of living by growing irrigated crops or by protecting their rain-fed crops from the vagaries of Monsoon rainfall. It is also their only source for drinking water for the family and cattle. In many developing countries, like in India, hard rock hydrogeologists have, therefore, an important role to play.

Fig. 2. Red bole (inter-trappean bed) sandwiched between hard, fractured basalt flows. (Exposure seen in road-side cutting on Pune-Bangalore highway Pune District, Maharashtra State.)

2. Occurrence of ground water

Ground water under phreatic condition occurs in the soft mantle of weathered rock, alluvium and laterite overlying the hard rock. Under this soft mantle, ground water is mostly in semi-confined state in the fissures, fractures, cracks, and joints. (Deolankar 1980) In basaltic terrain the lava flow junctions and red boles sandwiched between two layers of lava flows, also provide additional porosity (Fig 2). The ratio of the volume of water stored under semi-confined condition within the body of the hard rock, to the volume of water in the overlying phreatic aquifer depends on local conditions in the mini-watershed. Dug-cum-bored wells tap water from the phreatic aquifer and also from the network of fissures, joints and fractures in the underlying hard rock. (Fig 3 A and Fig 3B).

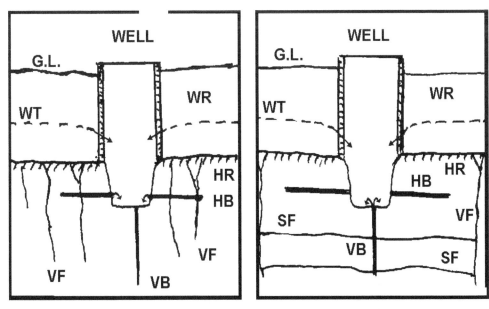

GL – Ground Level, HB – Horizontal Bore, HR – Hard Rock, SD – Sheet Fracture or joint,
VB – Vertical Bore, VF – Vertical Fracture, WR – Weathered rock, WT – Water Table.

Fig. 3A. and 3B. Dug cum Bored Wells

The recharge to ground water takes place during the rainy season through direct infiltration
into the soft mantle overlying the hard rock and also into the exposed portions of the
network of fissures and fractures. In India and other Asian countries in Monsoon climate,
the ratio of recharge to rainfall in hard rock terrain is assumed between 3 to 15%. (Limaye
S.D & Limaye D.G. 1986) This ratio depends upon the amount and nature of precipitation,
the nature and thickness of topsoil and weathered zone, type of vegetation, evaporation
from surface of wet soil, profile of underlying hard rock, the topographical features of the
sub-basin and the status of soil and water conservation activities adopted by villagers.
Ground water flow rarely occurs across the topographical water divides and each basin or
sub-basin can be treated as a separate hydrogeological unit for planning the development of
ground water resources. After the rainy season, the fully recharged hard rock aquifer
gradually loses its storage mainly due to pumpage and effluent drainage by streams and
rivers. The dry season flow of the streams is thus supported by ground water outflow. The
flow of ground water is from the peripheral portions of a sub-basin to the central-valley
portion, thereby causing de-watering of the portions closer to topographical water divides.
In many cases, the dug wells and bore wells yielding perennial supply of ground water can
only be located in the central valley portion.

The annual recharge during Monsoons being a sizable part of the total storage of the aquifer,
the whole system in a sub-basin or mini-basin, is very sensitive to the availability of this
recharge. A couple of drought years in succession could pose a serious problem. The low
permeability of hard rock aquifer is a redeeming feature under such conditions because it
makes small quantities of water available, at least for drinking purpose, in the dug wells or
bore wells in the central portion of a sub-basin. If the hard rocks had very high permeability,

the ground water body would have quickly moved towards the main river basin, thereby leaving the tributary sub-basins high and dry. The low permeability in the range of 0.05 to 1.0 meter per day thus helps in retarding the outflow and regulating the availability of water in individual farm wells. More farmers are thus able to dig or drill their wells and irrigate small plots of land without causing harmful mutual interference.

3. Ground water development

In the highly populated but economically backward areas in hard rock terrain, Governments in many developing countries have taken up schemes to encourage small farmers to dig or drill wells for small-scale irrigation. This is especially true for the semi-arid regions where surface water resources are meager. For example, in peninsular India, hard rocks such as granite, gneiss, schist, quartzite (800,000 sq kms) and basalts (Deccan traps- 500,000 sq kms) occupy about 1.30 million sq. kms area out of which about 40% is in semi-arid zone, receiving less than 750 mm rainfall per year. Over 4.00 million dug wells and bore wells are being used in the semi-arid region for irrigating small farm plots and for providing domestic water supply.

Development of ground water resources for irrigational and domestic use is thus a key factor in the economic thrift of vast stretches of semi-arid, hard rock areas. The basic need of millions of farmers in such areas is to obtain an assured supply for protective irrigation of at least one rain-fed crop per year and to have a protected, perennial drinking water supply within a reasonable walking distance. The hard-rock hydrogeologists in many developing countries have to meet this challenge to impart social and economic stability to the rural population, which otherwise migrates to the neighboring cities. The problem of rapid urbanization by exodus of rural population towards the cities, which is common for many developing countries, can only be solved by providing assurance of at least one crop and rural employment on farms.

Ground water development in a sub-basin results in increased pumpage and lowering of the water table due to the new wells, resulting in the reduction of the effluent drainage from the sub-basin. Such development in several sub-basins draining into the main river of the region reduces the surface flow and the underflow of the river, thereby affecting the function of the surface water schemes depending on the river flow. In order to minimize such interference, it is advisable to augment ground water recharge by adopting artificial recharge techniques during rainy season and also during dry season. The measures for artificial recharge during Monsoon rains include contour trenching on hill-slopes, contour bunding of farms, gully plugging, farm-ponds, underground stream bunds, and forestation of barren lands with suitable varieties of grass, bushes and trees. Artificial recharge in dry season is achieved through construction of percolation tanks.

However, increase in pumpage takes place through the initiative of individual farmers to improve their living standard through irrigation of high value crops, while recharge augmentation is traditionally considered as Government's responsibility and always lags far behind the increase in pumpage. In many parts of the world, particularly in developing countries, groundwater is thus being massively over-abstracted. This is resulting in falling water levels and declining well yields; land subsidence; intrusion of salt water into freshwater supplies; and ecological damages, such as, drying out wetlands.

Groundwater governance through regulations has been attempted without much success, because the farmers have a strong sense of ownership of ground water occurring in their farms. Integrated Water Resources Development (IWRM) is being promoted as a policy or a principle at national and international levels but in practice at field level, it cannot be attained without cooperation of rural community. NGOs sometimes play an important role in educating the villagers and ensure their cooperation.

4. Importance of dry season recharge

During the rainy season from June to September the recharge from rainfall causes recuperation of water table in a sub-basin from its minimum level in early June to its maximum level in late September. This is represented by the equation:

$$P = R + ET + r$$

Where P is the precipitation, R is surface runoff, ET is evapo-transpiration during the rainy season and r is the net recharge, represented by the difference between the Minimum storage and Maximum storage in the aquifer. However, after the aquifer gets fully saturated, the additional infiltration during the Monsoons is rejected and appears as delayed runoff. During the dry season, depletion of the aquifer storage in a sub-basin, from its maximum value to minimum value, is represented by the following equation:

(Aquifer storage at the end of rainy season i.e. Maximum storage) =

(Aquifer storage at the end of summer season, i.e. Minimum storage) +

(Pumpage, mainly for irrigation, during the dry season from dug wells & bore wells) +

(Dry season stream flow and underflow supported by ground water) –

(Recharge, if any, available during the dry season, including the return flow from irrigated crops)

The left-hand side of the above equation has an upper limit, as mentioned above. On the right-hand side, the minimum storage cannot be depleted beyond a certain limit, due to requirement for drinking water for people and cattle. Dry season stream flow and underflow supported by ground water have to be protected, as explained earlier, so that the projects depending upon the surface flow of the main river are not adversely affected. Any increase in the pumpage for irrigation during dry season due to new wells must therefore be balanced by increasing the dry season recharge.

The best way to provide dry season recharge is to create small storages at various places in the basin by bunding gullies and streams for storing runoff during the rainy season and allowing it to percolate gradually during the first few months of the dry season. Such storages created behind earthen bunds put across small streams are popularly known as percolation tanks. (Fig. 4 and Fig 5). In semi-arid regions, an ideal percolation tank with a catchment area of 10 sq. kms. or so, holds maximum quantity by end of September and allows it to percolate for next 4 to 5 months of winter season. Excess of runoff water received in Monsoon flows over the masonry waste weir constructed at one end of the earthen bund. By February or March the tank is dry, so that the shallow water body is not exposed to high rates of evaporation in summer months. (Fig.6) Ground water movement being very slow, whatever quantity percolates between October and March, is available in the wells on the downstream side of the tank even in summer months till June or the beginning of next

Monsoon season. Irrigation of small plots by farmers creates greenery in otherwise barren landscape of the watershed. (Fig.7). Studies carried out in granite-gneiss terrain have indicated that about 30% of the stored water in the tank percolates as recharge to ground water in the dry season. The efficiency is thus 30%. In basaltic terrain, if the tank is located at suitable site and the cut-off trench in the foundation of tank-bund does not reach up to the hard rock, higher efficiencies up to 70% could be obtained. (Limaye D.G & Limaye S D. 1986) However, more research is required for estimation of the impact of percolation tanks in recharge augmentation. In the state of Maharashtra in western India, over 10,000 percolation tanks have been constructed so far. (DIRD website, 2011) They are beneficial to the farmers and are very popular with them.

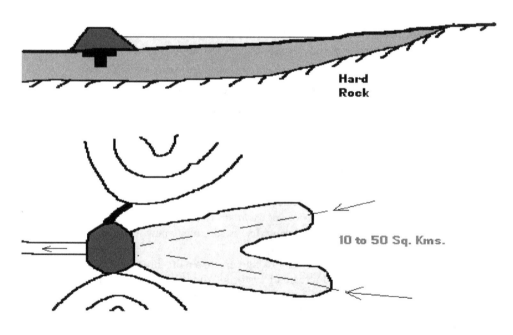

Fig. 4. Cross section and plan of a typical percolation tank on a stream between two hillocks. Hillocks are represented by black contours. Earthen Bund of the tank is shown in brown. Cut-off trench below the bund is in black and accumulated rain water is in pale blue color.

Fig. 5. Stone Pitching on the face of the earthen bund of a percolation tank under construction. Photo from village Hivre Bazar, District: Nagar, Maharashtra state.

Fig. 6. A percolation tank about to get dry towards beginning of summer. Location: Village: Ralegan Siddhi, District: Nagar, Maharashtra State.

Storage Bund

Percolation Tank

A Farm Pond.

Open Well

Fig. 7. Greenery crated within a dry, semi-arid watershed with the help of water conservation, farm ponds and percolation tank. The open well has shallow water table even in summer. A low cost centrifugal pump would soon be installed for small scale irrigation. (Location: Village Hivre Bazar. Dist: Nagar, Maharashtra State)

The initial efficiency of a percolation tank reduces due to silting of its bottom by receiving muddy runoff from the watershed. If the watershed is well-forested and has a cover of grass, bushes and crops, the silting is minimal. But in an average of 5 to 6 Monsoon seasons the tank bed accumulates about 0.20 to 1.00 meters of silt. Silt reduces the storage capacity of the tank and also impedes the rate of vertical flow of recharge because of its low permeability. The efficiency gets reduced due to silting and de-silting of tank bed when it dries in summer, becomes necessary. (Limaye S D. 2010).

Another type of recharge available during the dry season is the return flow or the percolation below the root zone of crops, from irrigated farms. This return flow to ground water is usually estimated at about 25% to 30% of the volume of ground water pumped in dry season and applied for irrigation. However, due to increasing popularity of more efficient irrigation methods like sprinkler or drip systems, this type of recharge has a declining trend.

5. Conclusions

A watershed is the meeting point of climatology & hydrology. It is therefore, necessary to manage our watersheds so as to absorb the climatic shocks likely to come from the erratic climatic patterns expected in near future. This can be done only through practicing soil and water conservation techniques for artificial recharge during rainy season and through construction of small percolation tanks for artificial recharge during the dry season.

Basin or Sub-Basin management begins with soil and water conservation activities taken up with people's active participation in several sub-basins within a large basin. This improves the shape of hydrograph of the stream or river in the basin, from a 'small time-based and sharp-peaked hydrograph' to a 'broad time-based and low-peak hydrograph'. Such a change also increases ground water recharge.

Small water storages or tanks created in the sub-basins by bunding streams and gullies, store runoff water during the Monsoon season and cause recharge to ground water during the next few months of dry season. The residence time of water in the basins is thus increased from a few months to a few years and the percolated water is available in the wells even during the summer season of a drought year.

After a few years of operation, silting of the tank bed reduces the volume of water stored and also the rate of vertical infiltration. Regular desilting of tanks by local people is, therefore, advisable.

A national policy for afforestation of degraded basins with proper species of grass, bushes and trees should be formulated. Afforestation with eucalyptus trees should not be encouraged in low rainfall areas as this effectively reduces ground water recharge. The main aim of forestation of a degraded watershed with local spices of hardy trees, grasses etc. should be to conserve soil, reduce velocity of runoff water, promote recharge to ground water and increase the biomass output of the watershed. .

Involvement of NGOs should be encouraged in forestation schemes and soil & water conservation programs so as to ensure active participation of rural community in recharge augmentation. NGOs also motivate the farmers to maintain the soil and water conservation structures put in by Government Departments so as to ensure long-term augmentation of recharge to ground water. Along with such management on supply side, demand management is also equally important. NGOs play a significant role in promoting the use of efficient irrigation methods and selection of crops with low water requirement.

The website www.igcp-grownet.org of the UNESCO-IUGS-IGCP Project GROWNET (Ground Water Network for Best Practices in Ground Water Management in Low-Income countries) gives several best practices including soil and water conservation, recharge augmentation etc. for sustainable development of ground water. The author of this Paper is the Project Leader of GROWNET. The reader is advised to visit the website for detailed information.

Although the discussion in the Paper refers to hard rock terrain in India, it would be equally applicable to many other developing countries, having a similar hydro-geological and climatic set-up.

6. References

Adyalkar PG and Mani VVS (1971) Paloegeography, Geomorphological setting and groundwater possibilities in Deccan Traps of western Maharashtra. Bull Volcanol. 35: pp 696-708

Deolankar S B (1980) The Deccan Basalts of Maharashtra State, India: Their potential as aquifers. Ground Water Vol. 18 (5): pp 434-437.

DIRD website www.dird-pune.gov.in/rp_PercolationTank.htm Efficiency of percolation tanks in Jeur sub Basin of Ahemadnagar District, Maharashtra state, India. (Visited June 2011)

Limaye D G & Limaye S D (1986) Lakes & Reservoirs: Research and Management 2001. Vol.6: pp 269-271

Limaye S D and Limaye D G (1986) Proc. of International Conference on ground water systems under stress. Australian Water Resources Conference Series 13. pp 277-282

Limaye S.D. (2010) Review: Groundwater development and management in the Deccan Traps (Basalts) of western India. Hydrogeology Journal (2010) 18: pp 543-558

5

Alternative Management Practices for Water Conservation in Dryland Farming: A Case Study in Bijar, Iran

Fardin Sadegh-Zadeh[1], Samsuri Abd Wahid[1],
Bahi J. Seh-Bardan[1], Espitman J. Seh-Bardan[2] and Alagie Bah[1]
[1]Department of Land Management, Faculty of Agriculture,
Universiti Putra Malaysia, Serdang, Selangor
[2]Department of Water Science, Faculty of Agriculture, Zabol University,
[1]Malaysia
[2]Iran

1. Introduction

1.1 Water conservation

Water conservation in the arid and semi arid regions is an important issue that influences both the environment and crop production. Runoff which is induced by rainfall can cause soil erosion which poses a dominant threat against long-term sustainability of farming (Derpsch et al., 1986). A further problem usually associated with runoff is the loss of soil particles that may pollute water bodies. Pollutants commonly found in runoff include soil particles, phosphorous, nitrogen, pesticides, etc. (Motavalli et al., 2003a)

During runoff, soil particles in the form of displaced sediments are carried along with the flowing water. The runoff mostly settles around the edge of a dam and the sediments it contains will subsequently be deposited underneath the reservoir. This continuous and gradual silting of the dam over time will consequently affect its capacity to store water. The decrease in the capacity of reservoir depends on the concentration of soil particles in the river that supplies the dam. In spite of decades of concerted research efforts, sedimentation is still considered the most serious problem threatening the dam industry. The deposition of soil particles in the dam may decrease the efficiency of the dams' turbines.

1.2 Soil and water conservation practices in dryland farming

Dryland farming is the profitable production of useful crops without irrigation on lands that receive annual rainfall of less than 500 mm per year. In the arid and semiarid regions, the conservation of precipitation water for crop production is very vital. In dryland crop production areas, a major challenge is to conserve precipitation water appropriately for use during crop growth (Baumhardt and Jones, 2002). It is imperative that farming practices should conserve and utilize the available rainfall efficiently. To optimize water storage under any precipitation condition, the soil should have enough infiltrability, permeability and capacity to store water. Water is the main constraint in dryland farming in the West of Iran. Precipitations tend to accrue during winter, while crops' growth season in spring is accompanied by high temperatures. These conditions are the constraints limiting crop production in dryland agriculture in Iran (Hemmat and Eskandari, 2004b).

1.3 Tillage

The objective of tillage operations is to improve soil conditions including porosity, temperature, and soil water storage capacity for increased crop production (Alvarez and Steinbach, 2009). Tillage systems that practise conservation farming during the winter are known as important methods in controlling soil erosion and runoff (Alvarez and Steinbach, 2009; Derpsch et al., 1986). Tillage practices play an important role in dry farming agriculture; however, the appropriate implements, their time and method of use have to be specific for different agro-climatic zones.

1.3.1 Conservation tillage

Conservation tillage research studies have focused on the effects of tillage practices on soil and moisture conservation for increased crop production, water conservation and soil erosion control. Several studies have attempted to develop appropriate and sustainable tillage and residue management methods that would maintain favorable soil conditions for crop growth. After harvest, stubble mulch is accumulated on the soil surface. Such materials do not only prevent direct impact of raindrops on the soil, but also impede the flow of water down the slope, thereby decrease the water flow on the soil surface and increase the amount of infiltration water (Hemmat et al., 2007). Conservation tillage systems have the potential to improve soil quality and reduce soil loss by providing protective crop residue on the soil surface and improving water conservation by decreasing evaporation losses (Su et al., 2007). Tillage creates a rough cloddy surface that lengthens the time necessary for the rain to break down the clods and seal the surface. Reduced tillage practices have been used in the production of many crops to increase soil water conservation (Locke and Bryson, 1997; Peterson et al., 1998). Reduced tillage practices protect soils from erosion and runoff by maintaining more crop residue on the soil surface. The magnitude and trends of change in soil physical properties depends on antecedent conditions, wheel tracks, soil texture and climate (Hobbs et al., 2008). However, contradictory results have been reported in literature about these effects. Mahboubi et al. (1993) showed the beneficial effects of long-term conservation tillage systems including chisel plowing and no-tillage compared to conventional tillage in ameliorating soil physical properties.

Compacted soils of arid regions have low organic matter contents and are proned to crusting which may decrease infiltration, seedling emergence and plant growth (Unger and Jones, 1998). For soils that are hard setting or have a root-restricting layer, some form of mechanical loosening through deep tillage is necessary to conserve the soil and water in order to facilitate crop growth (Nitant and Singh, 1995; Vittal et al., 1983). On the other hand in some soils, water conservation and water and wind erosion contros are major goals of conservation tillage systems. Any tillage method that keeps residue on the surface may protect the soil against dispersion by rain drop impact and the pounded or flowing water will decrease crusting (Hoogmoed and Stroosnijder, 1984; Pikul Jr and Zuzel, 1994)

1.3.2 Comparing various tillage systems

Studies have revealed that tillage operations do modify soil properties including soil structure (Roger-Estrade et al., 2004; Saggar et al., 2001), bulk density and porosity (GLSB and KULIG, 2008; Lampurlanés, 2003; Unger and Jones, 1998), water retention and distribution (Hemmat et al., 2007), root growth and yield (Box Jr and Langdale, 1984; Busscher and Bauer, 2003; Shirani et al., 2002; Su et al., 2007).

Conventional tillage methods used by farmers result in physical degradation of soil and increased soil erosion and runoff (Unger and Fulton, 1990). Excessive tillage results in deterioration in the soil environment and also increases the cost of production. On the other hand, the no- tillage system can affect the growth and establishment of plants through increased weed competition and poor soil physical conditions. Reduced tillage has been found to be feasible in improving soil properties (Locke and Bryson, 1997; Peterson et al., 1998). Each tillage system modifies soil properties differently. Moldboard plow buries plant residues and stubble, but chisel plow enables retention of plant residues on soil surface.

1.3.3 Organic amendments and tillage

Previous studies have reported that application of sewage sludge, compost, farmyard manure and other kinds of organic amendments resulted in a significant increase in aggregate stability, water content, hydraulic conductivity and infiltration and a decrease in bulk density (Arshad and Gill, 1997; Bahremand et al., 1999; Motavalli et al., 2003b; Shirani et al., 2002). Some literature reported that application of manure to the soil decreased soil compactibility (Mosaddeghi et al., 2003). They also showed that mixing manure with the soil does not only decrease compactibility but also decrease subsoil compaction.

1.4 Mulching and water conservation

Stubble mulching aims at disrupting the soil drying process by protecting the soil surface at all times either with a growing crop or with crop residues left on the surface during fallow. The first benefit of the stubble mulch is that wind speed is reduced at the surface by up to 99% and, therefore, losses due to evaporation are significantly reduced (GLSB and KULIG, 2008). In addition, crop residues can improve water infiltration (Hemmat and Taki, 2001) and decrease water runoff losses (Morin et al., 1984). Layered mulch could keep soil moist, change soil moisture regime and help to keep the soil moist(Sadegh-Zadeh et al., 2009). The decrease in evaporation by layered mulch was due to the ability of the mulch to decrease soil temperature during the hot-dry season. Other studies on mulching and soil moisture showed that tephra mulch could keep more soil moisture than the un-mulched soil and tephra mulch were able to change aridic soil moisture regime into a udic one (Diaz et al., 2005; Tejedor et al., 2002).

1.5 Justification of the study

Dryland production of wheat is the main cultivation system that accounts for the largest area of Iran (Hajabbasi and Hemmat, 2000; Hemmat; Hemmat and Eskandari, 2004a; Hemmat and Eskandari, 2006). In the semi-arid region of Iran, most of the precipitations occur in the late autumn, winter and early part of spring, while the growth of wheat is almost in the late spring. Hence, there water is not sufficient to grow wheat. On the other hand, most of the precipitation water are lost as runoff, particularly for bare lands and when conventional tillage systems are used (Hemmat and Eskandari, 2006).

Chisel plow enables retention of maximum amount of stubble and residues on the soil surface and there is no induced hard pan in soil profile (Barzegar et al., 2003). Consequently, there tends to be an increase in water infiltration and storage, leading to a decrease in soil erosion (Barzegar et al., 2003). Deep plowing with subsoiler has similar characteristics to that of the chisel plow, but the only difference is the plowing depth. Studies have shown that deep tillage system can improve soil physical properties including decreased bulk density, infiltration rate and hydraulic conductivity, increased soil moisture in soil profile and yield under dryland production (Busscher et al., 2000; Busscher et al., 2002; Busscher et al., 2006; Laddha and Totawat, 1997; Motavalli et al., 2003b; Nitant and Singh, 1995) .

Conservation tillage is the recommended method which helps to retain the crop residues in the soil surface at the same time conserves the soil and water (Sow et al., 1997). However, the presence of stubble and crop residues in soil surface may negatively influence yield (Hemmat and Taki, 2001). Hence, farmers seem not to prefer practicing conservation tillage. The use of moldboard plow is the frequently used method in this area but it buries stubble and plant residues and produce a hard pan in the bottom of plow layer (Barzegar et al., 2003).

Mulching is another feasible method to conserve water in semi-arid and arid regions (Sadegh-Zadeh et al., 2009). Considerable amount of literatures have been published on various tillage operations commonly used in some parts of Iran (Hemmat and Eskandari, 2004b; Hemmat and Eskandari, 2006; Mosaddeghi et al., 2009; Shirani et al., 2002). However, there is no reported study on the combined effects of tillage systems and mulching practices in the arid and semi-arid areas characterized by seasonal hot and dry summer and cold winter. The production of winter wheat (*Triticum aestivum* L.) of the cultivar Sardary is commonly practiced in this area.

1.6 Objectives of the study

The general objectives of the present study were to develop appropriate tillage and farm yard mulching systems for water and soil conservation in an aridisols with the aim of improving the grain yield of wheat (Triticum aestivum L.) of the cultivar Sardary. Specific objectives of the study were to (i) compare the effect of moldboard plow (MP), chisel plow (CP) and deep plow (DP) on soil properties, soil water content, runoff, soil loss and grain yield of wheat, (ii) investigate effect of farmyard mulch (FM) on soil properties, water content, runoff, soil loss and grain yield of wheat, and (iii) introduce a tillage system which is capable of conserving precipitation water to optimize grain yield and decease surface runoff and soil loss.

2. Materials and methods

2.1 Site characteristics and soils

The experiment was conducted at three sites in one of the famous areas of dryland wheat production in Bijar, Kurdestan province of Iran. The soils belong to the Aridisols order (Soil Survey Staff, 2006). The soil of the experimental sites consist of different textural classes (sandy loam, loam and clay loam). The mean annual precipitation of the region is 400 mm, most of which is received from late autumn, winter and early spring. During winter, most of the precipitation water is converted into snow. The soils of the region have low organic matter and nitrogen, with medium amount of phosphorus and high potassium content. The soil had been cultivated since long time ago. The climate of the area is characterized by a cold and snowy winter and a warm and dry summer with high evapotranspiration potential (in excess of 1500 mm in an evaporimetric tank).

2.2 Experimental site and design
2.2.1 Tillage treatments
Three tillage treatments were imposed during seed bed preparation. The plot layout was arranged using a randomized complete block design with four replicates. Plowing operations were carried out in April 2003 and disking was performed twice in September of 2003.

Tillage systems used were as follows:

Moldboard plow (MP) (200 mm depth) and twice offset disking (70 mm depth).

Chisel plow (CP) (300 mm depth) and twice offset disking (70 mm depth).

Deep plow (DP) with subsoiler (450 mm depth) and twice offset disking (70 mm depth).

The experimental design for each soil type was a spilt-split plot with three tillage systems as main plots, manure applications (no application, application of 3 mm thickness of farmyard manure (FM) on soil surface after sowing and mix same amount of FM with the soil surface (70 mm depth before sowing) as split plots, and planting (no planting and planting) as split–split plots. Each plot size was 2 m × 20 m in four replications.

Fertilizer including urea, ammonium phosphate, and microelements (Zn, Mn, Fe and Cu) were applied before sowing according to soil analysis results and recommendation rates. Wheat (*Triticum* aestivum L.) seeds (cultivar Sardary) were sown (at a rate of 150 kg ha-1) and weeding was done manually.

2.3 Runoff and soil loss
Runoff and soil loss were measured in each plot. The plot edges were made of solid materials (wood plank). The edges of the plots were about 15 cm above the soil surface to prevent input from splashes entering the plot from the surrounding areas and were sufficiently embedded into the soil. Runoff and soil loss were measured by collecting the runoff water in 40-liter capacity buckets (Khan and Ong, 1997), which were placed at the bottom of each plot. The collection buckets were connected to the runoff plots via PVC tubes, which collected both soil sediments and runoff water from the each plot after every rainfall event. Sediment concentration was determined through sampling collected runoff at the out let of each plot. Sediment content was determined by means of drying and weighing (Inbar and Llerena, 2000). Sediment yield was assumed to be equal to the rate of soil erosion. Runoff and sediment measurements were conducted from cultivation to harvesting stages.

2.4 Measurement of soil properties
The measured soil properties were pH, $CaCO_3$ content, soil water content at field capacity (FC) and permanent wilting point (PWP), organic matter (OM) content, particle size distribution, electrical conductivity of saturation extract (ECe), cation exchange capacity (CEC), and soil bulk density. Soil bulk density was measured on undisturbed

core samples (Blake and Hartge, 1986). Particle size distribution was determined by the Bouyoucos hydrometer method (Bouyoucos, 1962). Water infiltration rates were determined in the soil surface of various treatments using a double-ring infiltrometer (Bouwer, 1986). The CEC was determined according to method used for alkaline soils (Bower et al., 1952). The pH and electrical conductivity were determined from a saturated paste extract (Rhoades, 1982). The amount of $CaCO_3$ was determined by acid neutralization method (Allison and Moodie, 1965) and the OM content was determined by the potassium dichromate oxidation method (Nelson et al., 1982). The soil water content was measured using gravimetric method. Water retention capacity was measured at FC (− 33 kPa) and PWP (− 1500 kPa) (Gardner and Klute, 1986). Soil water content was measured at depths of 1 to 100 cm in every 5 cm intervals by the gravimetric method.

Wet aggregate stability was determined using the method of Kemper and Rosenau (Kemper and Rosenau, 1986). Fifty grams of air-dried aggregates (3–5 mm diameter) from each soil type was wet sieved through a 2 mm sieve. The sieving time was 10 min at 50 oscillations per minute. The percent of aggregate size bigger than 2 mm was calculated and used as an aggregate stability index among treatments. Soil compaction was determined using the Cone index readings which were taken with a hand held 13-mm diameter, 30 ° cone tip penetrometer (Carter, 1967) at soil surface of each plot. The soils were sampled to determine their properties during the months of October (2003), April (2004) and June (2004) to represent the planting time, middle and end of wheat growth, respectively. The dry weight of roots per plot was measured at harvest.

3. Result and discussions

The soil properties of the experimental sites are shown in Table 1. There were considerable differences between the various soils in term of soil pH, $CaCO_3$, FC, PWP, texture, CEC, and slope. Results of sandy loam and clay loam soils were similar to loam soil; therefore, only result of loam soil is presented in this chapter. Figure 1 shows the location of the experimental site which is adjacent to a watershed areasituated behind the Golbolagh reservoir dam.

Soils	Slope (%)	FC	PWP	$CaCO_3$	OM	EC_e (dS m^{-1})	pH	CEC (cmol$_c$ kg^{-1})
		\(g kg^{-1})						
Sandy loam	4	210	95	130	14	1.1	7.1	11
Loam	8	270	107	140	13	0.8	7.5	15
Clay loam	5	300	114	90	16	0.7	7.3	18

FC: field capacity; PWP: permanent wilting point; ECe: electrical conductivity of saturation extract; CEC: cation exchange capacity; OM: organic matter.

Table 1. The properties of the soils

Fig. 1. Location of experimental site (a) and location of site in the watershed behind the Golbolagh dam (b).

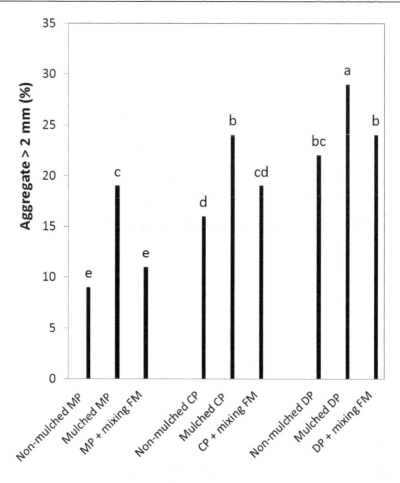

Treatments

Fig. 2. The percentage of aggregates bigger than 2 mm among treatments in April (2004).
Values followed by the same letter are not significantly different (P < 0.05).

3.1 Aggregate stability

Soil aggregate stability can be evaluated by determiningthe % of aggregates bigger than 2 mm
(Hajabbasi and Hemmat, 2000). The soil aggregates bigger than 2 mm at different treatments
are shown in Figure 2. There was significant difference (P < 0.05) in soil aggregates percentage
among treatments. Non-moldboard plow (MP) had the lowest aggregate percentage. Mixing
farmyard manure (FM) with soil increased the aggregate percentage but was not significantly
different (P < 0.05). Application of the FM as mulch on soil surface enhanced the aggregate
percentage significantly. The percentage of aggregates bigger than 2 mm in non-mulched
chisel plow (CP) was higher than non-mulched MP. The application of FM as mulch in CP
increased the percentage of aggregates significantly (P < 0.05). Mixing the FM with soil in the
CP increased aggregates percentage but it was not statistically significant. The percentage of

aggregate bigger than 2 mm in deep plow (DP) was more than MP and CP. Mulched DP had highest percentage of aggregate. Mixing of FM with DP soil increase the aggregate percentage compared to non-mulched DP but it was not significant.

Application of FM as a mulch on all kind of plowing increased the percentage of soil aggregates in soil surface. This result is in agreement with result of Shirani et al. (2002) who showed that the mixing of 30 and 60 Mg ha[-1] of FM increased the aggregate stability. However, there is no reported data on the application of FM as a mulch and aggregate stability. As mentioned in the materials and methods section, the thickness of FM mulch was 3 cm and if we calculate the weight of FM per ha, it is around 5 Mg ha[-1]. Although the amount of FM applied was low, it increased the aggregate stability drastically and improved the soil structure.

3.2 Soil cone index

Soil compaction is normally determined by measuring its penetration resistance with a penetrometer and the value obtained is referred to as a soil cone index. The soil cone indices of the treatments are shown in Figure 3. There was significant difference (P < 0.05) in soil cone index among treatments. The soil cone indexes in mulched treatments were much lower than either non-amended treatments or treatments of FM mixed with the soil. Lowest cone index was observed in mulched-DP and non-mulched MP which had the highest cone index. Mixing FM with the soil decreased soil cone index compare to the same tillage without application of FM. The percentage of soil aggregate bigger than 2 mm was negatively correlated with the cone index (Figure 4.). Soil cone index decreased with increasing amount of aggregates bigger than 2 mm.

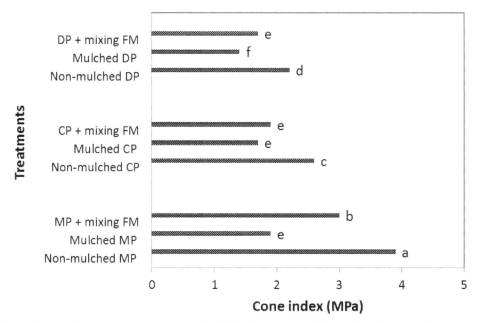

Fig. 3. Cone index of treatments in April (2004). Treatments followed by the same letter are not significantly different (P < 0.05).

Soil crust appeared when soil aggregates are broken down into smaller particles (Bissonnais, 1996; Le Bissonnais et al., 1989). The high cone index values observed in non-mulched MP could be due to the restriction factor for wheat emergence and water infiltration; hence, higher soil cone index could be a potential limiting factor for plant growth and it is expected that precipitation water will be lost as surface runoff.

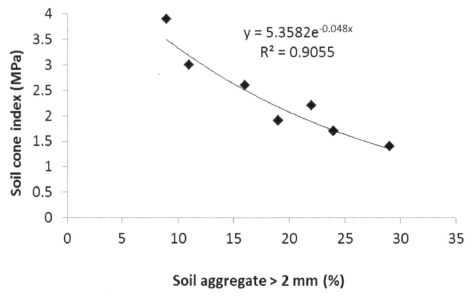

Fig. 4. Correlation between percentage of soil aggregate bigger than 2 mm and soil cone index.

3.3 Soil bulk density

The soil bulk densities measured at different depths of treatments are shown in Figure 5. There was no difference in soil bulk density between non-mulched MP, mulched MP and mixed FM with MP treatments at depths lower than 10 cm. However, mulched MP had the lowest bulk density at the soil surface (5 cm depth). Mixing FM with soil decreased the soil bulk density at the upper depths (5 and 10 cm depths) compared to same non-mulched MP treatment. The highest bulk density in non-mulched MP was observed at 25 cm depth and this is possibly due to the presence of hardpan at that depth. The bulk densities in the MP treatments increased at depths lower than 20 cm. This is reasonable because MP is able to loosen the upper 20 cm soil layer and below this depth, the soil can be compacted by moldboard during tillage operation. Chisel plow treatments had lower bulk density at 25 cm depth in contrast to MP treatments. This data shows that hardpan was broken by the CP operation. Bulk density was lowest in the DP treatments. This plowing method decreased soil bulk density atall depths except at 50 cm. The addition of FM as mulch helped to lower the surface soil bulk density in all tillage systems. This result is consistent with the finding of Lampurlanés, (2003) who showed that deep tillage could keep soil to be porous. This result also in agreement with Shirani et al. (2002) who reported that farmyard manure significantly decreased soil bulk density on the row tracks of field.

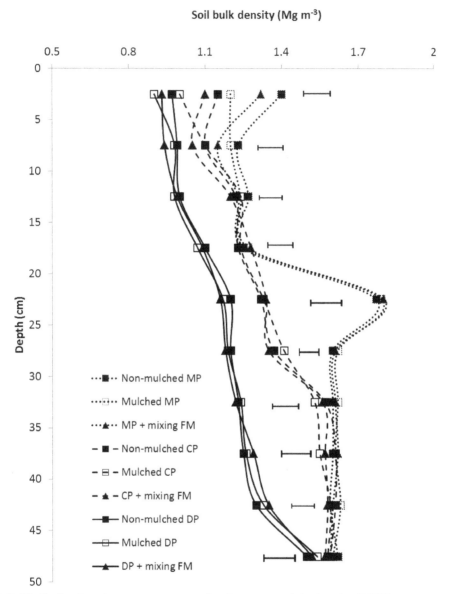

Fig. 5. Soil bulk density of treatments versus depths measured during April 2004. Horizontal bars represent the common LSD (P < 0.05) by depth for all treatments.

3.4 Infiltration rate

The infiltration rates of the treatments are shown in Figure 6. The infiltration rate in mulched DP was the highest, while non-mulched MP had lowest rate. The infiltration rates of the treatments were in the following order: mulched DP > mulched CP > DP + mixing FM

= mulched MP > non-mulched DP = CP + mixing FM > non-mulched CP > MP + mixing FM > non-mulched-MP. The results indicate that infiltration rates increased with the application of mulch on soil surface in all tillage systems. Among the tillage systems studied, the DP which has the lowest bulk density and cone index has the highest infiltration rate followed by CP and MP. Application of FM as mulch in the all tillage treatments increased infiltration rate. Mixing FM with soil increased infiltration rate but it was not as effective as the FM mulch. This indicated that the addition of FM as a mulch helped improved soil structure and increase water infiltration. This can be attributed to the fact that FM mulch can increase aggregate stability, increase soil water content and decrease runoff. This finding is in accordance with the results of Shirani et al. (2002), which showed that mixing 30 and 60 Mg ha^{-1} of farmyard manure increased soil hydraulic conductivity. However, there is no reported study on the application of FM as mulch on soil surface in various tillage systems after sowing. Only 3 to 4Mg ha^{-1} of FM is required to cover the soil for FM to be used as a mulch which is affordable to the farmers in the large area of dryland production of Iran. This application method is also suitable for dry land production in other parts of world.

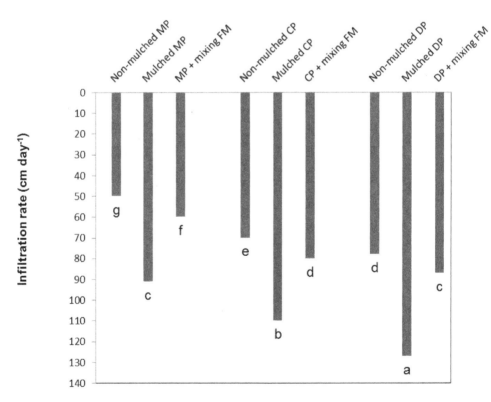

Fig. 6. Infiltration rate of the treatments during April (2004). Treatments followed by the same letter are not significantly different (P < 0.05).

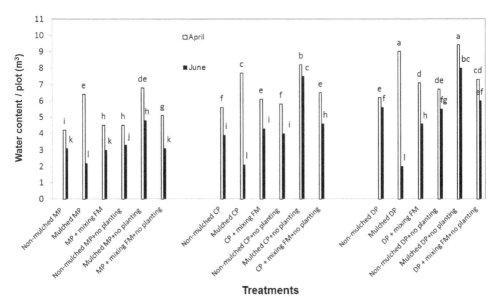

Treatments

Fig. 7. Soil water content of treatments in April and June (2004). Treatments followed by the same letter are not significantly different (P < 0.05).

3.5 Soil water content

Effects of mulch and tillage treatments on water content in the soil profile, evaporation, and availability of water for plant growth could not be studied in winter and autumn because biomass production started in the April and plants needed water since beginning of April to end of June. Soil water contents of treatments during April and June (2004) are shown in Figure 7. There was a significant difference in soil water content among MP, CP and DP in April. Soil water content was significantly (P < 0.05) higher in DP than CP and MP during April. The lowest water content was observed in MP treated soils in April. There was higher water content in soils treated with mulch compared to same treatment without mulch. Soil water content of non-cultivated plot of DP with mulch was the highest among the tillage treatments. The soil water content of non-cultivated plot of CP with mulch was lower than the non-cultivated DP with mulch. However, non-cultivated plot of CP with mulch had higher water content than non-cultivated plot of MP with mulch. The water content in all planted treatments which had mulch was lowest. Higher biomass in the mulch treated plots implied the higher water consumption in these treatments. Conversely, in the non-planted mulch treatments for all tillage operations (MP, CP and DP) water content were higher than same tillage systems without mulch. The differences in water content in the non-planted treatments with planted treatments could be due to reduced transpiration in the non-planted treatments. The water contents of all treatment during June were decreased in comparison to April. The decrease in water content could be ascribed to evaporation and possible drainage of water to a depth greater than 100 cm. Our data suggest that FM mulch on soil surface was effective in conservation of water and thus retaining more water in the soil than that in the other treatments. This result is in agreement with previous result which showed the addition of   municipal sewage sludge increased available water in the root zone mainly due to reduction in evaporation (Agassi et al., 2004).

3.6 Runoff

Annual runoff for each treatment is presented in Table 2. Almost all precipitations occurred in autumn 2003 and winter 2004 and then continued in April 2004. Precipitations were extremely low after April 2004. Lack of rainfall in May and June (2004), i.e., during the plant growth period, was considered as a vital aspect of water conservation and runoff control in this area during the previous autumn and winter. Spreading FM mulch on the soils resulted in decreased runoff of MP, CP and DP compared to the same tillage systems without mulch. For example, in the case of MP, mulching could decrease 300 m^3 ha^{-1} of runoff per year compared to the non-mulched MP.

The amount of runoff in the CP and MP treatments was negligible and it was zero in the mulched CP and DP (Table 2). Addition of FM as a mulch decreased the runoff of CP and DP compared to the non-mulched CP and DP (Table 2). Therefore, it can be concluded that the decrease in runoff is one of beneficial effects of FM mulch especially in the conventional tillage system (MP). A decrease in runoff was related to the higher infiltration rate in mulched treatments and this finding is in agreement with other studies which showed that mulching was effective in controlling runoff in soils susceptible to sealing (Poesen and Lavee, 1991; Saxton et al., 1981).

3.7 Soil loss

The trend of soil loss was similar to the trend of runoff among the treatments (Table 2). Soil loss ranged from 0 kg $plot^{-1}year^{-1}$ for mulched CP and MP to 193 kg $plot^{-1}year^{-1}$ for non-mulched MP. (Table 2). Less soil loss was measured on mulched MP treatment compared to the non-mulched MP treatment. Mulching also decreased soil loss in CP and DP. However, in some cases the differences between the mulched and non-mulched treatments was not significant ($P < 0.05$). Overall, soil loss in all treatments of CP and DP was negligible. Mulching significantly decreased the soil loss in some treatments by preventing the impacts of raindrops on the soil aggregates (Figure 2) hence conserving the soil structure (Figure 3) and as a result infiltration rates were higher in mulched treatments compared to the non-mulched treatments (Figure 6). From theresults it can be concluded that mulching was very useful to control soil erosion. This result is in agreement with the findings of Döring et al. (2005) who reported that soil erosion was reduced by more than 97% in a rain simulation experiment on a potato field of 8% slope with 20% crop cover compared to sils without crop cover.

3.8 Wheat emergence, dry weight of roots and grain yield

Wheat emergence was significantly influenced by mulching and tillage (Table 2). Wheat emergence in April (2004) was low under tillage systems without mulch. Mulching increased the wheat emergence in all tillage systems (Table 2). The highest and the lowest rates of wheat emergence were observed in mulched MP and non-mulched DP, respectively. Although conservation tillage (non-mulched CP and non-mulched DP) were able to decrease water and soil loss as runoff and sediment and prevented soil structure from being degradation, there was less wheat emergence compared to the conventional tillage (non-mulched MP). The decrease in wheat emergence rate in non-mulched CP and DP could be due to cracking of seeds at the end of autumn when the water freeze into ice . However, mulching seems to decrease seed cracking in the mulched-CP and DP treatments. A possible explanation for a decrease in wheat emergence in non-mulched MP is crusting was induced by raindrops in this treatment.

Effects of tillage systems and mulching on dry weight of roots and grain yield were significant (Table 2). Weight of roots and grain yield under all the tillage systems generally increased with the application of mulch. Mulched MP had higher weight of roots and grain yield than the non-mulched MP which could be explained by the higher wheat emergence and higher water content for feeding the plant during the growth period.

The grain yield was the highest in the mulched DP. Mulched CP had lower grain yield than the mulched MP. Although non-mulched DP had high water content, the grain yield and weight of roots were low probably because the absence of enough wheat emergences (Table 2). The existence of enough moisture in the soil profile (Figure 7) is not a sufficient condition to have high amounts of yield. High wheat emergence is also necessary to have enough plant population and economical grain yield. In the non-mulched CP and DP there was not enough plant population to obtain an economical yield. This is the reason why farmers do not prefer conservation tillage in this area. However, the addition of FM as mulch to the tillage systems increased the yield drastically (approximately two folds) and solved the problem of low grain yield often associated with the conservation tillage. The FM mulch helped the soils to store enough water in the soil profile but at the same time did not reduce the wheat emergence. Therefore, the FM mulch method is highly recommended for conserving water in the arid and semi-arid regions without jeopardizing the grain yield of wheat.

	Treatments								
	Non-mulched MP	Mulched MP	MP + mixing FM	Non-mulched CP	Mulched CP	CP + mixing FM	Non-mulched DP	Mulched DP	DP + mixing FM
Runoff (m³/ plot year)	1.8 a†	0.6 c	1.3 b	0.2 d	0 e	0.15 de	0.14 de	0 e	0.10 de
Soil loss (kg / plot year)	193 a	53 c	114 b	67 c	0 e	29 d	59 c	0 e	11 e
Wheat emergence per plot	7600 d	12072 a	9480 b	6037 f	8360 c	7320 d	5722 g	9317b	6681e
Dry weight of roots (kg / plot)	6.96 e	10.26 c	7.99 d	6.79 e	11.89 b	7.19 de	7.57 d	15.63 a	7.90 d
Grain yield (kg / plot)	4.87 g	8.78 c	6.40 e	5.93 ef	11.95 b	7.61 d	6.30 de	15.28 a	7.24 d

†Means in a row followed by a different letter differ significantly based on the LSD at $P < 0.05$.

Table 2. Runoff, soil loss, wheat emergence, dry weight of roots and grain yield in various treatments

4. Conclusions

The objective of this study was to develop appropriate tillage and farm yard mulching systems for conserving water and soil with the aim of improving the grain yield of wheat (Triticum aestivum L.) of the cultivar Sardary. Three plowing treatments (MP, CP, and DP) and three FM applications (no application, mixing with soil and application as mulch on soil surface) were employed. From the results it can be concluded that:

- Conservation tillage systems (CP and DP) increased soil water content in the soil profile and decrease runoff and soil loss. However, the yield was not economical due to the effect of ice damage on the winter wheat seed.
- Application of FM as a mulch to the conservation tillage systems increased grain yield. The mulched DP had the highest yield of wheat among the treatments.
- Mulching increased infiltration, soil water content and yield in the conventional tillage system (MP).
- Mulching enhanced the wheat yield in all tillage systems and at the same time conserves water and soil. Therefore, it is a good strategy to be adopted not only with the conventional tillage system but also with the conservation tillage system which is usually associated with low yield.

5. References

Agassi M., Levy G., Hadas A., Benyamini Y., Zhevelev H., Fizik E., Gotessman M., Sasson N. (2004) Mulching with composted municipal solid wastes in Central Negev, Israel: I. effects on minimizing rainwater losses and on hazards to the environment. Soil and tillage research 78:103-113.

Allison L., Moodie C. (1965) Carbonate volumetric calcimeter method. Methods of analysis. Agronomy Monogr:1389-1392.

Alvarez R., Steinbach H. (2009) A review of the effects of tillage systems on some soil physical properties, water content, nitrate availability and crops yield in the Argentine Pampas. Soil and tillage research 104:1-15.

Arshad M., Gill K. (1997) Barley, canola and wheat production under different tillage-fallow-green manure combinations on a clay soil in a cold, semiarid climate. Soil and tillage research 43:263-275.

Bahremand M., Afyuni M., Hajabbasi M., Rezainejad Y. (1999) Short and mid-term effects of organic fertilizers on some soil physical properties. pp. 288-289.

Barzegar A., Asoodar M., Khadish A., Hashemi A., Herbert S. (2003) Soil physical characteristics and chickpea yield responses to tillage treatments. Soil and tillage research 71:49-57.

Baumhardt R., Jones O. (2002) Residue management and tillage effects on soil-water storage and grain yield of dryland wheat and sorghum for a clay loam in Texas. Soil and tillage research 68:71-82.

Bissonnais Y. (1996) Aggregate stability and assessment of soil crustability and erodibility: I. Theory and methodology. European Journal of Soil Science 47:425-437.

Blake G.R., Hartge K. (1986) Bulk density. Agronomy (USA).

Bouwer H. (1986) Intake rate: Cylinder infiltrometer. Agronomy.

Bouyoucos G.J. (1962) Hydrometer Method Improved for Making Particle Size Analyses of Soils1. Agronomy Journal 54:464.

Bower C., Reitemeier R., Fireman M. (1952) Exchangeable cation analysis of saline and alkali soils. Soil science 73:251.

Box Jr J.E., Langdale G. (1984) The effects of in-row subsoil tillage and soil water on corn yields in the Southeastern coastal plain of the United States. Soil and tillage research 4:67-78.

Busscher W., Bauer P. (2003) Soil strength, cotton root growth and lint yield in a southeastern USA coastal loamy sand. Soil and tillage research 74:151-159.

Busscher W., Frederick J., Bauer P. (2000) Timing effects of deep tillage on penetration resistance and wheat and soybean yield.

Busscher W., Bauer P., Frederick J. (2002) Recompaction of a coastal loamy sand after deep tillage as a function of subsequent cumulative rainfall. Soil and tillage research 68:49-57.

Busscher W., Bauer P., Frederick J. (2006) Deep tillage management for high strength southeastern USA Coastal Plain soils. Soil and tillage research 85:178-185.

Carter L.M. (1967) Portable penetrometer measures soil strength profiles. Agric. Eng 48:348-349.

Derpsch R., Sidiras N., Roth C. (1986) Results of studies made from 1977 to 1984 to control erosion by cover crops and no-tillage techniques in Paraná, Brazil. Soil and tillage research 8:253-263.

Diaz F., Jimenez C., Tejedor M. (2005) Influence of the thickness and grain size of tephra mulch on soil water evaporation. Agricultural Water Management 74:47-55.

Döring T.F., Brandt M., Heß J., Finckh M.R., Saucke H. (2005) Effects of straw mulch on soil nitrate dynamics, weeds, yield and soil erosion in organically grown potatoes. Field crops research 94:238-249.

Gardner W.H., Klute A. (1986) Water content. Methods of soil analysis. Part 1. Physical and mineralogical methods:493-544.

Glsb T., Kulig B.. (2008) Effect of mulch and tillage system on soil porosity under wheat (Triticum aestivum). Soil & tillage research 99:169-178.

Hajabbasi M., Hemmat A. (2000) Tillage impacts on aggregate stability and crop productivity in a clay-loam soil in central Iran. Soil and tillage research 56:205-212.

Hemmat A., Taki O. (2001) Grain yield of irrigated winter wheat as affected by stubble-tillage management and seeding rates in central Iran. Soil and tillage research 63:57-64.

Hemmat A., Eskandari I. (2004a) Conservation tillage practices for winter wheat-fallow farming in the temperate continental climate of northwestern Iran. Field crops research 89:123-133.

Hemmat A., Eskandari I. (2004b) Tillage system effects upon productivity of a dryland winter wheat-chickpea rotation in the northwest region of Iran. Soil and tillage research 78:69-81.

Hemmat A., Eskandari I. (2006) Dryland winter wheat response to conservation tillage in a continuous cropping system in northwestern Iran. Soil and tillage research 86:99-109.

Hemmat A., Ahmadi I., Masoumi A. (2007) Water infiltration and clod size distribution as influenced by ploughshare type, soil water content and ploughing depth. Biosystems engineering 97:257-266.

Hobbs P.R., Sayre K., Gupta R. (2008) The role of conservation agriculture in sustainable agriculture. Philosophical Transactions of the Royal Society B: Biological Sciences 363:543.

Hoogmoed W., Stroosnijder L. (1984) Crust formation on sandy soils in the Sahel I. Rainfall and infiltration. Soil and tillage research 4:5-23.

Inbar M., Llerena C.A. (2000) Erosion processes in high mountain agricultural terraces in Peru. Mountain Research and Development 20:72-79.

Kemper W., Rosenau R. (1986) Aggregate stability and size distribution.

Khan A.A.H., Ong C.K. (1997) Design and calibration of tipping bucket system for field runoff and sediment quantification. Journal of soil and water conservation 52:437.

Laddha K., Totawat K. (1997) Effects of deep tillage under rainfed agriculture on production of sorghum (Sorghum biocolor L. Moench) intercropped with green gram (Vigna radiata L. Wilczek) in western India. Soil and tillage research 43:241-250.

Lampurlanés J. (2003) Soil bulk density and penetration resistance under different tillage and crop management systems and their relationship with barley root growth. Agronomy Journal 95:526.

Le Bissonnais Y., Bruand A., Jamagne M. (1989) Laboratory experimental study of soil crusting: Relation between aggregate breakdown mechanisms and crust stucture. Catena 16:377-392.

Locke M.A., Bryson C.T. (1997) Herbicide-soil interactions in reduced tillage and plant residue management systems. Weed Science:307-320.

Mahboubi A., Lal R., Faussey N. (1993) Twenty-eight years of tillage effects on two soils in Ohio. Soil Science Society of America journal (USA).

Morin J., Rawitz E., Hoogmoed W., Benyamini Y. (1984) Tillage practices for soil and water conservation in the semi-arid zone III. Runoff modeling as a tool for conservation tillage design. Soil and tillage research 4:215-224.

Mosaddeghi M., Mahboubi A., Safadoust A. (2009) Short-term effects of tillage and manure on some soil physical properties and maize root growth in a sandy loam soil in western Iran. Soil and tillage research 104:173-179.

Mosaddeghi M., Hemmat A., Hajabbasi M., Alexandrou A. (2003) Pre-compression stress and its relation with the physical and mechanical properties of a structurally unstable soil in central Iran. Soil and tillage research 70:53-64.

Motavalli P., Anderson S., Pengthamkeerati P. (2003a) Surface compaction and poultry litter effects on corn growth, nitrogen availability, and physical properties of a claypan soil. Field crops research 84:303-318.

Motavalli P., Stevens W., Hartwig G. (2003b) Remediation of subsoil compaction and compaction effects on corn N availability by deep tillage and application of poultry manure in a sandy-textured soil. Soil and tillage research 71:121-131.

Nelson D., Sommers L., Page A. (1982) Methods of soil analysis. Methods of soil analysis 9.

Nitant H., Singh P. (1995) Effects of deep tillage on dryland production of redgram (Cajanus cajan L.) in central India. Soil and tillage research 34:17-26.

Peterson G., Halvorson A., Havlin J., Jones O., Lyon D., Tanaka D. (1998) Reduced tillage and increasing cropping intensity in the Great Plains conserves soil C. Soil and tillage research 47:207-218.

Pikul Jr J., Zuzel J. (1994) Soil crusting and water infiltration affected by long-term tillage and residue management. Soil Science Society of America (USA).

Poesen J., Lavee H. (1991) Effects of size and incorporation of synthetic mulch on runoff and sediment yield from interrils in a laboratory study with simulated rainfall. Soil and tillage research 21:209-223.

Rhoades J. (1982) Soluble salts. Methods of soil analysis. Part 2:167-179.

Roger-Estrade J., Richard G., Caneill J., Boizard H., Coquet Y., Defossez P., Manichon H. (2004) Morphological characterisation of soil structure in tilled fields: from a diagnosis method to the modelling of structural changes over time. Soil and tillage research 79:33-49.

Sadegh-Zadeh F., Seh-Bardan B.J., Samsuri A.W., Mohammadi A., Chorom M., Yazdani G.A. (2009) Saline Soil Reclamation By Means of Layered Mulch. Arid Land Research and Management 23:127-136.

Saggar S., Newman R., Ross C., Dando J., Shepherd T. (2001) Tillage-induced changes to soil structure and organic carbon fractions in New Zealand soils. Australian Journal of Soil Research 39:465-489.

Saxton K., McCool D., Papendick R. (1981) Slot mulch for runoff and erosion control. Journal of soil and water conservation 36:44.

Shirani H., Hajabbasi M., Afyuni M., Hemmat A. (2002) Effects of farmyard manure and tillage systems on soil physical properties and corn yield in central Iran. Soil and tillage research 68:101-108.

Su Z., Zhang J., Wu W., Cai D., Lv J., Jiang G., Huang J., Gao J., Hartmann R., Gabriels D. (2007) Effects of conservation tillage practices on winter wheat water-use efficiency and crop yield on the Loess Plateau, China. Agricultural Water Management 87:307-314.

Tejedor M., Jiménez C., Díaz F. (2002) Soil moisture regime changes in tephra-mulched soils: implications for soil taxonomy. Soil Science Society of America Journal 66:202-206.

Unger P.W., Fulton L.J. (1990) Conventional-and no-tillage effects on upper root zone soil conditions. Soil and tillage research 16:337-344.

Unger P.W., Jones O.R. (1998) Long-term tillage and cropping systems affect bulk density and penetration resistance of soil cropped to dryland wheat and grain sorghum. Soil and tillage research 45:39-57.

Vittal K., Vijayalakshmi K., Rao U. (1983) Effect of deep tillage on dryland crop production in red soils of India. Soil and tillage research 3:377-384.

6

Economic Principles for Water Conservation Tariffs and Incentives

John P. Hoehn
Michigan State University
United States of America

1. Introduction

Water conservation creates no water. It manages water and water scarcity. Water conservation shifts water and water scarcity across people, their water uses, space and time. Water is scarce when it is insufficient to satisfy all the valued uses that different people have for water. Valued uses include water for drinking, cleaning, industry, transporting waste, recreation, and sustaining environmental goods such as habitat, ecosystem and aesthetic services.

Water scarcity is most obvious in droughts (Kallis, 2008), but scarcity is routine even where water appears physically abundant. Water is scarce in Chicago, Illinois, even though it lies adjacent to a lake containing more than 1,180 cubic miles of water (Ipi & Bhagwat, 2002). Conflicts between people who want water for *in-situ* uses such water for recreation and ecological services and people who want water to withdraw water for people, agriculture and industry are common in both humid and arid environments (World Commission on Dams, 2000).

People manage water scarcity through any number of formal organizations and informal groupings. These organizations and groups are water management institutions. Legislation, law and regulation establish formal institutions. Formal institutions include municipal water agencies, water districts, corporations and local governments. Other institutions emerge informally out of customs, habits, histories and the politics of water problems. Informal institutions include urban water markets that arise in neighborhoods that are not served by a municipal network (Crane, 1994) and the patterns of priorities, rights and expectations that guide irrigation in traditional societies (Ostrom, 1990). Legislation and law often intervene to recognize, modify and transform informal institutions into formal ones (cf. Coman, 2011).

Different institutions have different effects on water conservation. Within one irrigation district, farmers may face 'use-it-or-lose-it' rules. Use-it-or-lose-it rules force farmers to use their water seasonal allocations in a given year or forfeit the unused portion (Spangler, 2004). In another district, rules may be set up so that farmers may leave unused allotments in a reservoir and stored for future use. The two irrigation districts may have the same consequences under normal conditions. When a prolonged drought occurs, farmers in the first district may watch their crops shrivel from water scarcity, while farmers in the second district draw on their banked water and enjoy a normal crop year.

Rules, fees, restrictions and institutional policies make some actions beneficial and others relatively costly. The relative benefits and costs of different actions are economic incentives.

Incentives encourage some behaviors and discourage others. Incentives may shape behavior in ways that are consistent with objectives but they can also lead to behavior that is entirely unexpected.

Municipal water systems use tariffs to collect the revenue necessary to sustain and expand a water system but some tariff choices inadvertently create incentives that weaken financial sustainability. For instance, municipal water systems often adopt water tariffs that supply a subsistence quantity of water for a payment that is less than the cost of provision. When small users predominate, provision below cost eventually makes service financially infeasible. Reliable service areas then shrink to service only higher income neighbors and the poor are left to purchase water at many times the highest fees charged by the water agency (Rogerson, 1996; Komives et al., 2005; Saleth & Dinar, 2001).

Conservation is effective when incentives are consistent with conservation goals. Economic analysis of incentives is part of integrated water management (Snellen & Schrevel, 2004). Economic principles help identify the relative values of water in different uses and set up processes to balance water uses in ways that are consistent with its scarcity value and conservation goals. Analysis of benefits and costs is an inherent part of sustainable investments. The water resource investments required to satisfy the thirsts of cities and towns or irrigate agriculture cannot be sustained without the careful financial management of benefits, costs, revenues and expenditures.

The objectives of this chapter are to identify the economic principles central to water resource management and to examine how these principles are used in the process of designing water conservation tariffs and incentives. Tariffs are the pricing mechanisms used by municipal water agencies to raise revenues from water use. The analysis examines how tariffs may be structured, set and implemented to provide incentives for efficient water conservation.

The primary economic principles are opportunity cost, demand, deadweight loss, trade, and third-party effects. Opportunity costs are the building blocks of economic cost and valuation. Opportunity cost is not a physical or accounting concept. Opportunity cost is a relative value concept based upon the value of a resource in its next best use. It is the value forgone by using a resource in a particular way rather than in its next most valuable use. Opportunity cost may be higher or lower than the value of a resource in its current use. When the opportunity cost of water is higher than its value in a current use, water is wasted.

User demands are the sources of water value and deadweight loss is a measure of value lost in the misallocation and waste of water. Demand is a relationship between a user's willingness to pay for an additional unit of water and the quantity of water available to that user. Demand value is a marginal or incremental concept; it measures the amount a user is willing to pay for the last unit of water consumed or used. For example, a thirsty person finds the first few sips of water highly valuable, but as a person's thirst is satisfied, additional swallows are successively less valuable. Deadweight loss combines demand values and opportunity cost to define an economic index of water waste.

Trade is a response of economic agents, people and firms, to a wasteful allocation of water. Water is wasted when water remains in low value uses while high valued uses go without. Economic agents find ways to trade and move water to higher valued uses when it physically possible and when they are empowered by law and custom to take ownership of the value of water. Trade requires physical infrastructure and the ownership and entitlement rules that support trade. Lack of infrastructure and mismatched rules and institutions are barriers to trade, standing in the way of an efficient allocation of water.

Third-party effects occur when upstream or downstream water users are not taken into account in water-use decisions. Water flows and water qualities connect different users in complicated and sometimes unforeseen ways. An upstream use of water may affect the quantity or quality of water available to downstream users. Water withdrawals by municipalities may reduce instream flows for recreation and ecological services. These unintended and unaccounted impacts are third-party effects.

The analysis is developed in the following way. The five primary economic principles are first defined and discussed. The subsequent section applies the economic principles to the design of efficient water conservation tariffs and to the evaluation of inefficient tariffs. The next section evaluates the tariff structures and tariff levels that are in use by municipalities around the world. The analysis indicates that very much remains to be done. Municipal systems contain large reservoirs of wasted water, reservoirs waiting to be tapped by efficient water conservation policies. The analysis concludes with three strategies to implement efficient water conservation incentives in residential water systems.

2. Five economic principles in water conservation

Water is a scarce resource. Economic scarcity means that there is not enough water available to meet all the wants and needs that people have for water. Economic scarcity is defined in reference to people's needs and wants rather than to physical availability. Needs and wants are defined broadly, to include the environmental and ecological services that make life possible and, so often, enjoyable. With all scarce goods, some wants and needs are unmet.

Scarcity makes water valuable. The values that people place on water make water worthy of considerable attention. When water is well-managed, water values enable the large investments necessary to ensure that essential values are protected and less essential values are supported with suitable quantities of water. When water is poorly managed, critical values are ignored and water is wasted in uses with little or no value.

Economic principles play a role in understanding and measuring water values. These principles make it possible to develop and evaluate water conservation incentives. At times, analysis of water values and incentives is highly technical and nuanced. The economic principles developed below are the basic concepts used to evaluate economic incentives and the decisions they motivate.

2.1 Opportunity cost

Using scarce water always has a cost. The scarcity of water means that there is always some other way the water may be used — some next best use. The cost is the value of the water in its next best use. The value forgone in the next best use is the opportunity cost of water. Opportunity cost is the fundamental principle of economic cost.

Opportunity cost varies across time and space. Time is important since water uses vary in quality, type and value over time. The values of water in agriculture rise and fall as seasons and growing conditions change. In winter, agricultural water values may be close to zero. Irrigation values rise substantially during the growing season, and especially during a drought. Water for outdoor recreation may show similar seasonal patterns. Water values also vary across space. Inability to transfer water across space due to lack of infrastructure or to legal barriers causes water values to diverge spatially. Divergent values are an incentive for human action to move water from a low value location to a high value location.

Divergent water values can lead to epic-scale investments in political power, litigation and infrastructure (Libecap, 2007).

Opportunity cost varies also with the quantity of water considered. The first unit of water transferred to the next best use has the highest value. Subsequent units transferred to the next best use have successively lower values. Marginal opportunity cost is the value of transferring a particular unit of water from its current use to its next best use. Marginal opportunity cost tends to fall as successive units of water are transferred from the current use to the next best opportunity.

2.2 Demand

Water demand is a relationship between water quantities and the amount users are willing to pay per-unit of water. The law of demand says that the amount a user is willing to pay per-unit declines as the amounts purchased increase. This means that there is an inverse relationship between willingness to pay and the amount of water available for use.

Household water use illustrates the law of demand. A small amount of water is highly valuable since it satisfies basic needs such as thirst and personal hygiene. Additional water for cooking and cleaning also has a high value, but not quite as high as the first few units of water used for drinking and hygiene. Household water values decline much further for values associated with gardening and lawn irrigation. Too much water may have negative values for a household – a leaky pipe may flood a basement and too much irrigation may destroy a productive agricultural field.

Water demand is represented mathematically with quantity as a function of price. Water demand for the ith water user is a function $w_i = f(t, \alpha)$ where w_i is a quantity of water demanded at price or volumetric charge, t, $f(\cdot)$ is the demand function and α represents other factors beside the volumetric charge that shift quantity demanded. The law of demand means that quantity demanded declines as the volumetric charge increases, so $\frac{dw_i}{dt} = \frac{df}{dt} < 0$.

Demand shifters, α, include variables such as user income, user age, seasons, weather, capital investments such as housing and acreage, water-use technology, regulatory restrictions, and information campaigns encouraging water conservation (Worthington & Hoffman, 2008). Households with greater incomes may use more water due to using more water-using appliances, larger gardens and lawns, swimming pools and other such uses. Water demand may shift seasonally since irrigation of gardens and lawns is more valuable in dry seasons than in wet seasons. Other factors that shift demand may include house and yard size, installation of water-saving technology, and knowledge of water saving strategies. Such demand shifters are the focus of non-tariff approaches to water conservation.

Water demands are estimated for a wide range of users, uses and aggregates of users and uses. Demands relevant to water conservation include household demands, crop demands, farm demands, industry demands, instream use demands and aggregates thereof, such as urban, agricultural and industrial demands. A common element is each of the latter demands is the law of demand, the inverse relationship between value as measured by willingness to pay and water quantity.

The law of demand is central to water conservation tariffs and incentives. The law of demand indicates that as volumetric tariff charges increase, the quantity of water demanded declines. Users adjust their water use downward in response to a volumetric charge increase. Users reduce their water use until the value they place on the last unit of water used or consumed is equal to the volumetric charge.

The opposite behavior happens with a reduction in a volumetric charge. A reduction in a volumetric charge means that the value that a user places on water exceeds the volumetric charge and the user responds by increasing water use. Water use increases until the user's valuation of the last unit of water is once again equal to the volumetric charge.

The responsiveness of demand to changes in a volumetric charge is summarized with a number called 'elasticity'. Elasticities are numbers that describe the percentage change in water use resulting from a one percent change in the volumetric charge. Elasticities are negative due to the law of demand. Estimated elasticities for residential water use tend to lie in a range from -0.3 to -0.6 with some reports of -0.1 or less (Dalhuisen et al., 2003; Nauges & Whittington, 2010; Worthington & Hoffman, 2008). An elasticity -.4 implies that water use declines by 4% for a 10% increase in a volumetric charge and by 40% for a 100% increase in a volumetric charge.

Elasticities are also estimated for demand shifters, α, and especially for the income levels of residential users. Income elasticities are useful in understanding how water use is likely to change with growth in incomes and with changes in the mix of income groups within service areas. An income elasticity of .4 means that annual growth in income of 4% is likely to increase water use by 1.2%. If such income growth continues over a decade, incomes rise by 34% and water use by 13.6%.

There are two important ranges of demand elasticities. Demand response is *inelastic* when a one-percent change in a volumetric charge or a shifter results in less than a one-percent change in water use. Demand response is *elastic* when a one percent change in price or a shifter results in a greater than one-percent change in water use. Residential water demands tend to be inelastic with respect to both volumetric charge and income (Dalhuisen et al., 2003).

2.3 Deadweight Loss

Deadweight loss is an economic measure of waste. Water is wasted when its value in a current use is less than its opportunity cost. Deadweight loss is the difference between current use value and opportunity cost when opportunity cost exceeds current use value.

Figure 1 illustrates deadweight loss with a simple case where a fixed amount of water is allocated between two users, person A and person B. The length of the horizontal axis represents the total amount of water available for use, 100 units. Water can be allocated to either A or B. Water allocated to A, Q_A, leaves 100 units minus Q_A, for B's use so $Q_B = 100 - Q_A$. At the left-hand corner of the diagram, A gets zero units of water and B gets 100 units. Moving from left to right along the axis, A gets more water and B gets less until A receives 100% of the water and B gets 0% at the right-hand corner of the figure.

A's demand curve is D_A. D_A slopes downward from left to right since A's value of the last unit of water consumed declines as A uses more and more water. Conversely, B's demand curve slopes upward from left to right as B gets less and less water. B's valuation of the last unit of water increases as B gets less and less water.

Water is wasted when its value in a current use is less than its opportunity cost. This means that water is wasted when A gets all the water since A's demand curve—the values that A places on successive units of water--lies below B's demand curve when A's allocation exceeds 55 units. The triangular area between the two demand curves from 55 to 100 units of water is the value forgone by giving A all the water. The triangle area is the deadweight loss of the allocation.

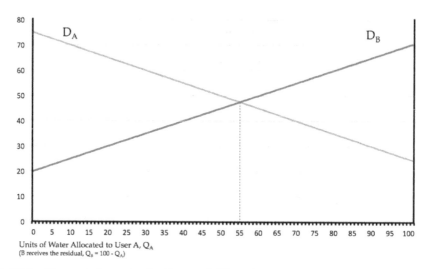

Fig. 1. Water Demand, Opportunity Cost and Allocation

Deadweight loss is the potential benefit of reducing A's use so that B can use more. For instance, by reallocating 45 units to B, the entire deadweight loss triangle from A's overuse is eliminated. When A uses 55 units and B uses 45 units, the demand values for the last unit of water used by each party are equal. Once the demand values are equal, there is no additional gain to letting B use more water. Letting B use more water than 45 units moves into the region where B's demand values are lower than A's.

Deadweight loss, wasted water, and inefficiency also result from allocating all water to B's use. At the lower left-hand corner of the the Figure 1, A gets no water and A's demand value exceeds B's demand value. Moreover, A's demand values exceed B's for all units of water up to A's use of 55 units and B's use of $45 = 100 - 55$ units. When A uses 55 units and B uses 45 units demand values for the last unit of use are again equal. The deadweight loss of B using all the water is the triangular area between the demand curves from 0 to 55 units. Having A use more than 55 units would move the allocation into a region of deadweight loss, where B's values for water exceed A's values.

There is no wasted water when there is no way to reallocate water use and improve the values associated with the allocation. Economic waste of water is zero only where the demand values are equal. In Figure 1, demand values are equal where A uses 55 units and B uses 45 units. At the latter allocation, zero water is wasted since current use exceeds opportunity cost and there is no deadweight loss.

Economics defines zero economic waste as an *efficient* allocation. An allocation that is not efficient is *inefficient*. An inefficient allocation wastes water and results in a non-zero deadweight loss.

Water conservation seeks to reduce waste and improve the efficiency of water use. A reduction in wasted water creates benefits by reducing deadweight loss and improving economic efficiency. A situation is fully efficient when opportunity cost is less than or equal to the current use value for all water uses. Full efficiency with zero waste and zero deadweight loss is unlikely in practice, but research shows that there are many practicable ways to reduce waste and improve efficiency.

2.4 Water trading

Water waste and inefficiency create a powerful economic incentive to reallocate and conserve water. For all the inefficient and wasteful allocations in Figure 1, the value of the last unit of water used is less than the value of an additional unit of water in the forgone use. For instance, when an allocation favors A with 100 units of water use, the value to B for a single unit of water exceeds the loss to A of giving up that single unit. A and B have an incentive to trade water for money or water. Trading isn't strictly in terms of water and money. Any good could stand in for money as long as it is valued and can be transferred to the ownership of the party that gives up a little water.

Starting from an allocation where A uses all the water, A and B can realize mutual gains if they voluntarily transfer a portion of A's water from A to B. If A is altruistic and gains value equivalent to B's value from merely knowing that B has water, A can simply give B some water. A second possibility is for B to compensate A by paying A for the loss of water. A and B can trade water for an amount of money somewhere between B's high value and A's low value. Trading at an intermediate value creates mutual benefits for both A and B. A trade of one unit of water from A to B eliminates the deadweight loss incurred through A's low valued use of that unit of water.

A and B have an incentive to continue trading water as long as there is a deadweight loss and a potential mutual benefit. By voluntarily continuing to trade, A and B eventually arrive at the efficient allocation of water shown in Figure 1 where A uses 55 units and B uses 45 units. A and B have the same incentives to trade when they begin with B using 100 units of water. In each case they trade to the efficient allocation where the demand values are equal, A uses 55 units, and B uses 45 units. Voluntary trading away from the efficient use allocation is not possible since once at the efficient allocation, opportunity cost is less than a user's demand value.

Reduction in water waste through voluntary trading is often difficult to achieve. In many situations, water customs, water rights law and lack of physical infrastructure make trade impractical or impossible (Slaughter, 2009). Trade in water requires a form of ownership consistent with trading. A buyer expects a transfer of a legal right to hold and use the water. Defining and implementing tradable ownership rights is often a slow and difficult process (Allan, 2003).

Trade in water also requires a water resource infrastructure. Water is physically heavy and difficult to transfer from one place and time to another. Water transfers require physical transport and storage facilities. These facilities become more complicated and costly with the complexity and scale of spatial and temporal transfers.

Water trading also requires an institutional infrastructure to identify water resources, to account for their location in space and time, and to define and enforce rules and procedures. A crucial economic feature of such trading rules and procedures is the degree that they distribute or consolidate resource ownership. Mistaken efforts in 'privatization' consolidate water treatment and distribution systems in a single owner. Single owners are all too likely to exploit their position as monopolists by restricting water access, raising water prices and increasing inefficiency and waste.

The cost and difficulty of developing efficient water trading infrastructure limits the practicability of water trading in many situations. Trade seems most feasible in dry regions around the world where water is particularly scarce, the opportunity cost of waste is high and the costs of physical transfer are relatively low (Grafton et al., 2010; Ruml, 2005).

2.5 Third-party effects

Water use and conservation involves decisions about how, when and where water is used. Third-party effects arise when such decision directly affect water availability to services and people that are not directly involved in a decision. Third-party effects are also denoted as externalities and spillovers and are relatively common in water management (Slaughter, 2009)

Water withdrawals from a water body potentially affect other water users. Water withdrawn from a reservoir for municipal use and irrigation may have negative impacts on boating, fishing and valued ecosystems. If so, these are negative third-party effects on boaters, fishers and those who value the ecosystem services. Negative third-party effects can also arise from irrigation drainage, from the toxins and pathogens in municipal and industrial wastewater, and from ground and surface water depleted by overuse.

Third-party effects may also be positive and beneficial. Construction of a reservoir funded by irrigators and municipal users may have positive third-party effects on boaters, fishers and ecosystem services. Treatment of urban wastewater may result in a recyclable product for irrigation and industrial cooling.

3. Efficient water conservation tariffs

Municipal water tariffs are the rates, charges and fees that municipal water systems charge users for water provided. Municipal tariffs are different from the prices that emerge from large markets. Market prices are typically the result of many buyers and sellers negotiating trades over time. Municipal water tariffs are usually set in an administrative and political setting. Administrative tariffs may be highly durable and may reflect political pressures more than the opportunity costs of the resources, including water, that are used in water treatment and distribution. Water is wasted and financial sustainability is threatened when tariff revenues do not cover costs.

Municipal water tariffs support water conservation to the extent that they encourage efficient water use and discourage waste. At the same time, efficient tariffs do not encourage overinvestment in water conservation and hoarding. Water conserving tariffs are just high enough to recover the economic costs of water provision, including the opportunity cost of water withdrawn from other uses and wastewater returned from the water system to the hydrological cycle.

3.1 Efficient water tariffs

Efficient water conservation tariffs have three attributes. First, they are simple enough that they can be accurately communicated to and understood by water users. Water users may not know how much they are charged for water use and only a subset of those who do know can work through the details of how reduced water use may save them money (Whitcomb, 2005). Rates can be complex and confusing even for an informed user (Martins et al., 2007; Dziegielewski et al., 2004).

Water tariffs need to be simple enough so that users can see how they can save money by reduced water use, careful conservation and investment in water saving technology. Gaudin (2006) finds that less than 20% of water utilities inform water users of the tariff schedule in water bills. When tariff details are clearly communicated and explained, water use falls by an average of 30% (Gaudin, 2006).

Second, water conservation tariffs provide the revenue necessary to cover the economic costs of water provision. This means, in part, that water conservation prices bring in enough revenue to pay the full financial costs of water provision, including the capital, maintenance, operating and administrative costs.

The financial opportunity costs of municipal water provision may be broken down into two components. The first is a fixed cost component, ϕ, that equals the financial costs of establishing a water and wastewater infrastructure of a given capacity. Fixed cost includes the capital investments costs of reservoirs, diversions, pipelines and treatment plants. The scale of the latter investments tends to be relatively fixed by their design capacity and varies relatively little with water volumes within the design capacities. Fixed financial cost also includes other costs that remain fixed within broad volume intervals. The latter include the overhead cost of an administration, accounting, and billing insofar as these do not vary with volumes processed. Fixed cost may be adjusted to account for new capital investments (Griffin, 2001).

The second portion of financial opportunity cost varies with the volume of water and wastewater processed. Variable costs arise from the labor, equipment, chemicals and energy required to treat, distribute, and maintain service quality and reliability as larger volumes of water and wastewater are processed. The variable cost component is denoted, δw. Variable cost increases proportionately with total water use within the system, $w = \sum_{i=1}^{N} w_i$, where N is the number of water users, $i = 1, ..., N$. The factor of proportionality, δ, is the financial opportunity cost of providing an additional unit of water and wastewater services. It is the marginal financial opportunity cost of water.

The total financial cost, $f(w)$, is the sum of fixed and variable costs, $f(w) = \phi + \delta w$. Total financial opportunity cost is the market cost of capital and purchased inputs used in processing municipal water and wastewater services. They are 'financial' in a sense that they show up explicitly as expenditures in a municipal system's financial accounts. When a municipal system purchases water inputs and pays to eliminate wastewater impacts, the financial costs shows in the system's accounts. However, explicit payments for raw water and pollution impacts are often not made. In the latter case, raw water and wastewater incur an unpaid opportunity cost.

The third attribute of an efficient water conservation tariff is that it accounts the non-financial opportunity costs of raw water inputs and wastewater outputs. These non-financial costs may have a fixed component associated with the ecological and environmental services forgone due to investments such as reservoirs and pipelines. In an efficient tariff, these non-financial fixed costs are added into ϕ along with financial opportunity costs.

The greater share of opportunity cost is likely to vary with the quantity of water provided and wastewater returned to the hydrological system. Variable opportunity cost includes unpaid values of raw water when raw water would have otherwise been used in some other economic activity such as agriculture. Additional sources of potential opportunity costs are forgone instream uses, changes in ambient water quality due to wastewater effluents, and forgone future use when current use depletes future supplies. The latter opportunity cost arises in the case of reservoirs and groundwater reserves when increases in current use significantly increase future scarcity.

Opportunity costs that vary with the volume of water and wastewater are denoted λw. The factor of proportionality, λ, indicates how opportunity cost increases with an additional unit of water and wastewater services; it is the marginal opportunity cost of water and wastewater provision. The full economic cost, $e(w)$, of water and wastewater services is the sum of financial and non-financial opportunity costs, $e(w) = f(w) + \lambda w = \phi + \delta w + \lambda w$. The economic cost has two components, a fixed cost, ϕ, and a variable cost, $\delta w + \lambda w = (\delta + \lambda)w$. Fixed and variable economic costs are sums of financial and non-financial opportunity cost. The sum is made explicit in the formulation because variable cost turns out to be central to water conservation incentives. The sum of the two variable cost parameters, $\delta + \lambda$, is the full economic cost of providing an additional unit of water within a municipal system of a given capacity; it is the marginal economic cost of providing processed water and wastewater.

A water conservation tariff is efficient in the sense that it encourages no wasted water. An efficient tariff communicates the full economic cost of water and wastewater services. Since economic costs have fixed and variable components, an efficient tariff reflects both components: the variable cost of water and wastewater services provided to a user and the user's share of fixed cost (Coase, 1946).

The first component of an efficient tariff is a volumetric charge. An efficient tariff levees a volumetric charge, τ, equal to the economic cost of an additional unit of water and wastewater services. The efficient volumetric charge per unit of water is $\tau = \delta + \lambda$. For the delivery of w_i, the ith user pays $\tau w_i = (\delta + \lambda)w_i$. By charging τ per-unit to each user, a municipal system recovers the full variable economic cost, $(\delta + \lambda)w$, of providing N users with water and wastewater services, $w = \sum_{i=1}^{N} w_i$.

Empirical analysis indicates that users respond to volumetric charges by reducing water use as the charge per unit increases (Nataraj & Hanemann, 2011). An efficient volumetric charge, $\tau = \delta + \lambda$, presents water users with the full incremental economic cost of their water use decisions. As the law of demand indicates, a user's own valuation of water is initially large for the first few units of water. When a user's own valuation is greater than the efficient per-unit charge, τ, the user increases water use.

As water use increases, demand values decline by the law of demand. Through error or neglect, use may increase to the point where the user's own valuation is less than τ. When this shows up in a billing cycle, the user finds it worthwhile to cut back on water use to the point where the demand value is equal to the per-unit charge. This is an efficient level of water use where the demand value is equal to the marginal opportunity cost of water. An efficient volumetric charge eliminates wasted water.

An efficient volumetric charge also gives users an incentive to find and install water-saving technologies. When a technology saves water at a per-unit cost less than τ, the user benefits by installing the technology. Volumetric charges different from τ leads users to make wasteful decisions. A volumetric charge less than τ results in too much water use and too little investment in water-saving technologies. A charge greater than τ makes the opposite error: too little water is used and too much is invested in uneconomic water-saving technology.

The second component of an efficient tariff is a fixed charge. The fixed tariff component recovers the portion of the economic fixed cost, δ that is not covered by volumetric revenue, τw. A portion of the latter revenue covers the variable financial cost, δw that is paid to acquire labor and other resources necessary for operating the system. The second portion of volumetric revenue is the opportunity cost, λw, of resources used but not paid in a financial transaction.

Opportunity cost revenue, λw, reduces the economic fixed cost that needs to be supported by additional revenue. The amount of the net fixed cost, $\eta = \delta - \lambda w$, is positive when $\delta > \lambda w$ and zero, negative when $\delta < \lambda w$ and zero when $\delta = \lambda w$ (Hall 2009). When the net fixed cost is positive, the system requires an additional fixed charge to cover the remaining cost. When net fixed cost is negative, the system receives revenues in excess of its financial costs and may return a fixed rebate to water users. When the net fixed cost is zero, there is no need for a fixed charge.

Economic principles allow considerable leeway in determining how net fixed costs are allocated across users, though two constraints apply. Let η_i be the fixed charge to the ith water user. The first constraint is that users' fixed charges add up to the total net fixed cost, $\eta = \sum_{i=1}^{N} \eta_i$, where η_i is the fixed charge paid or fixed charge rebate received by the ith water user. The second constraint is that the η_i is unrelated to the volume of water used. When η_i is correlated with w_i then the fixed payment alters the way a user views the volumetric charge. Rather than viewing the volumetric charge solely in terms of τ, the user views the volumetric charge as higher or lower consistent with the degree of correlation with η_i and whether η_i is a payment or a rebate.

The fixed charge allows municipalities to address fairness and equity without altering the water conservation properties of an efficient volumetric charge. Fixed charge schedules might address fairness and equity with a variety of measures that are correlated with equity considerations such as income, but not directly correlated with adjustments in water use. Baberan and Arbues (2009) suggest household size as a factor. Other possible measures include η_i differentiated by class of user such as industrial, commercial and residential; by zoning and land use categories; by interior areas of homes; or by neighborhood development vintage.

An efficient water conservation tariff is composed of a fixed and volumetric charge, $\eta_i + \tau w_i$. The fixed component provides the revenues required to (a) cover net fixed cost and (b) address fairness and equity. The volumetric charge, τ, is set to equal the marginal financial and non-financial opportunity costs of water and wastewater provision. The volumetric charge communicates efficient water conservation incentives to all water users.

3.2 Inefficient water tariffs

Municipal water systems adopt and maintain rate structures for a variety of reasons unrelated to water conservation. Common tariff structures include uniform volumetric rates without fixed charges, flat rates, decreasing block rates, increasing block rates and different combinations of volumetric, flat, decreasing block and increasing block rates. Except for combined flat and volumetric rates, each of these alternative tariffs have a *structure* that discourages efficient water conservation and encourages inefficiency and waste.

A *uniform rate* without a fixed charge is a charge per-unit of service received by a water user. A uniform rate is a volumetric charge since the total amount paid by a user is the product of the per-unit charge and the volume of water used. With positive fixed costs, a uniform rate set to cover the full economic costs of water and wastewater use is greater than the efficient volumetric charge. Such a rate is too high for efficient water conservation. An excessive uniform rate causes users to forego water uses that are beneficial and wastes time, money and resources in inefficient water saving. A uniform rate equal to an efficient volumetric charge presents users with efficient incentives for water conservation, but fails to cover net fixed costs. Ignoring positive net fixed costs makes the system financially unsustainable, an all too common problem in municipal systems (Banerjee et al., 2008; Hoehn and Krieger, 2000; Organization for Economic Cooperation and Development [OECD], 2009).

A *flat rate* is a fixed charge per connection without a volumetric charge. Flat rates may be set to cover the economic cost of municipal water and wastewater. In systems without user metering, a flat rate is the only feasible alternative (OECD, 2009). The water conservation flaw in flat rates is that they place no cost on an additional unit of water. The user's cost of an additional unit of water is zero, so water is treated accordingly. Users make decisions accordingly, using water as if it is free rather than scarce and valuable.

Flat rates result in significant water waste and large economic costs. Users not only use water inefficiently, they also find it financially unwise to prevent 'unintentional' waste. Leaky valves go unrepaired and outdoor irrigation is left unmonitored — wastewater merits no attention when more can be obtained without cost. Moreover, much of the wasted water flows through the sewer and wastewater system, unnecessarily increasing wastewater treatment costs and the third-party costs of pollution and pollution-caused disease.

A *decreasing block rate* is a set of volumetric charges that decrease in a staircase fashion as water use increases. Levels of water use are divided into intervals called blocks that are the lengths of the steps. The height of a step is the volumetric charge. The highest volumetric charge is at the top of the staircase and volumetric charges decrease with each step or block as water use increases. A water user using enough water to cover two blocks pays two different rates for water use; one for the first block of water use and a lower rate for the second block. A user whose water quantity covers three blocks pays three different rates for water. The latter user pays the highest rate for the first block, an intermediate rate for the second, and the lowest rate for the third.

A decreasing block rate can cover economic costs, but it is does not encourage efficient water use and conservation. At most, no more than one of the blocks can have a volumetric rate consistent with efficient water conservation. The other blocks encourage too little or too much conservation. Oddly, the decreasing block structure gives individuals using the least amount of water the largest incentives for water conservation. Those using the most water face the weakest incentive for cutting back.

An *increasing block rate* is a set of volumetric charges that increase in a staircase fashion. The lowest charge occurs at the first block and the largest charge occurs at the last block. Like the decreasing block rate, an increasing block rate covering economic or financial cost is unlikely to send efficient water conservation signals to any block. Users at the first block face too small an incentive for water conservation and those at the last block invest too much in water conservation.

Increasing blocks are often adopted based on claims of fairness and equity. The claim rests on the idea that the poorest and most disadvantaged groups are likely to use the least water, so the initial low rate lowers the cost sustained by these users (OECD, 2009). Research indicates that the fairness and equity claim is not valid. Poor and disadvantaged users fail to benefit, even in cities where fairness and equity appear most needed. Increasing rates tend to be regressive for two reasons. First, the initial block rate is paid by all users, rich, poor and middle-income, so there is no relative gain to the poor. Second, poorly financed municipal systems often exclude the poor from water service, so the benefit of low rates goes entirely to the middle-income class and rich (Komives et al., 2005). The poor are all too often left outside the municipal system where water costs can be 2 to 60 times greater than municipal rates (Saleth & Dinar 2001).

The final category includes a wide-range of tariff structures that combine uniform, flat, decreasing and increasing rates in different ways for different user types. Tariffs can also be adjusted by seasons of the year such as summer or during droughts. The essential problem with these combined structures is that, like their component parts, they fail to present users with simple, understandable and correct incentives for efficient water conservation. The result of such mixed signals is wasted water with all its costs of unnecessary water and wastewater treatment, foregone beneficial uses, and ecological damages.

4. Tariff incentives for residential water conservation

Residential water tariffs are well recognized as a water conservation tool (State of California, 2008; Beecher et al., 2005). This section examines whether existing tariffs encourage water conservation. The section begins with a brief comparison of experience with tariff and non-tariff approaches to water conservation. Tariff structures and volumetric charges used by water systems around the world are then reviewed to determine the extent that existing tariffs are efficient. Most tariffs appear too low to incorporate non-financial opportunity costs.

4.1 Non-tariff tools for water conservation
Water managers often favor non-tariff tools for water conservation. Non-tariff approaches include informational campaigns, technology rebates, voluntary restrictions and mandatory restrictions accompanied by legal penalties. Research indicates that most informational campaigns and voluntary restrictions are unreliable as conservation tools (Olmstead & Stavins, 2009), though some well-structured informational campaigns may reduce water use by up to 8% (Renwick & Green, 2000). Rebates on efficiency toilets show no effect on water use (Renwick & Green, 2000).

Mandatory restrictions enforced with strong penalties can be effective where penalties are strictly enforced and violators are made to pay. In Aurora, Colorado, restrictions with penalties reduced summer water use by up to 26% (Kenny et al., 2008). Renwick and Green (2000) study California water systems serving 8 million people and find that restrictions with penalties reduce water use by 19 to 29%. However, a portion of the public vocally resists restrictions and fines. Imperfect monitoring, uneven enforcement and criminalization of civil behavior—such as caring for one's property—can result in public controversy (Atwood et al., 2007). Also, restrictions may reduce a targeted behavior, but they do nothing to encourage waste reduction in unrestricted uses.

Water conservation tariffs may also generate public resistance. No one likes a cost increase. Worse yet, however, is going without water service as millions do when inadequate tariffs fail to cover even financial costs (Nauges & Whittington 2009) or when excessive water withdrawals threaten instream recreational and environmental resources (Hickey & Diaz 1999).

The structure of efficient conservation tariffs allows costs to be distributed in ways that increase the degree of public acceptance. First, an efficient volumetric rate is fair since it addresses the full economic cost of using an additional unit of water. With an efficient rate, no one gets away by not paying the incremental cost of water. Second, the fixed charge can be adjusted to address equity concerns across users and to avoid putting excess burdens on those unable to pay.

4.2 Water tariff structures used by municipal systems

Efficient water conservation tariffs have both an efficient structure and an efficient level. The efficient structure has two parts, a volumetric charge and a fixed charge. The analysis examines data on tariffs to determine the extent that water tariffs in use diverge from an efficient structure.

A 2010 water tariff survey by Global Water Intelligence [GWI] describes water tariff structures and levels for 276 water systems worldwide. Table 1 lists the number of survey responses by region and the percentage distribution of five tariffs. Responses from Asian water systems comprise about one-third of the sample. European systems provide an additional third of responses. The remaining third of responses were obtained from water systems in Africa, the Middle East and North Africa (MENA), North America and South America, with North American systems providing about 12% of the responses.

Region[a]	Water Systems (#)	Regional Rate Structure Distribution (%)[a]					
		Vol.	Flat	I Block	D Block	Efficient	No Data
All	276	23	3	50	2	20	2
Africa	18	0	0	94	0	6	0
Asia	94	41	3	50	0	5	0
Europe	87	22	1	24	1	48	3
MENA	19	5	0	89	0	5	0
N. America	32	6	9	44	16	22	3
S. America	26	8	0	88	0	0	4

[a]"MENA" is the Middle East and North Africa, "N." is North and "S. is South. "Vol." means a volumetric rate, "Flat" means a fixed charge, "I Block" means an increasing block structure, and "D Block" means a decreasing rate structure.

Table 1. Municipal Tariff Structures

The most common tariff structure is the increasing block structure reported by 50% of the systems. The increasing block structure gives water users divergent and inefficient signals for water conservation. Ninety-four percent of systems in Africa use increasing block structures and more than 88% use these structures in the MENA and South America. The high incidence in less developed regions is unfortunate both for efficiency and equity. Wasted water erodes already low incomes and increasing block tariffs have the most regressive consequences for the poor (Komives et al., 2005). Notably, the increasing block structure is much less common in Europe and North America.

Efficient tariffs structures with volumetric and fixed charges are used in 20% of the systems surveyed. Almost half of the systems surveyed in Europe use efficient tariff structures. The high European incidence of efficient tariffs may reflect recent reforms reported by industry organizations (Beecher et al., 2005). Twenty-two percent of North American systems use an efficient structure. Efficient tariffs are least common in South America, MENA, Asia and Africa.

Volumetric charges alone are common in Asia and Europe, with 41% of Asian systems reporting volumetric rates. Volumetric rates can offer efficient incentives for water conservation, but to do so, tariff revenues are not likely to equal financial costs. A tariff based on an efficient volumetric rate alone risks financial insolvency. Flat and decreasing block structures are uncommon in the GWI data. The low incidence may reflect the nature of the sample. The sample is targeted to the largest systems worldwide and systems that are functioning adequately enough to respond to survey inquiries. Flat tariffs are the only alternative in the absence of water use metering and many water systems operate without such metering (Banerjee, 2008; World Health Organization & United Nations Children's Fund, 2000).

The incidence of tariff structures in smaller North American systems and towns also cautions extending the global survey results to all water systems. Beecher (2011) surveys 80 water systems in the north central area of the United States and finds that 44% use decreasing block tariffs, 18% use increasing block tariffs, and no systems use efficient structures. Dziegielewski et al., (2004) finds that 35% of 426 water systems in Illinois use decreasing block structures, 4% use increasing block structures and only 1% use an efficient structure. In 12 larger water systems in Colorado, 44% use either a volumetric charge or an increasing block tariff, 12% use decreasing block tariffs and none use an efficient tariff structure (Western Resource Advocates, 2004).

4.3 Water tariff levels set by municipal water systems

Efficient tariff levels are set so that the volumetric charge is equal to the financial and non-financial opportunity costs of providing an additional unit of water. Table 2 lists monthly average water rates based on the 2010 GWI survey. The second column in Table 2 lists income per capita within the systems responding to the survey. Overall, average income per capita is $20,595, but regional levels range from a low of $1,645 in Africa to a high of $48,119 in North America. The average monthly charge per 1,000 gallons is $4.53 for water and $3.32 for sewerage and wastewater. Sixty-four systems or almost 25% report no wastewater charge billed to water uses. The average combined water and wastewater charge is $7.08 per 1,000 gallons of water use.

Region[a]	Water Systems (#)	Income per Capita[b] ($)	Water and Wastewater Charge[d] ($ per 1,000 gallons)		
			Water	Wastewater	Water and Wastewater
All	276	20,595	4.53	3.32	7.08
Africa	18	1,645	2.09	0.70	2.79
Asia	94	12,736	2.63	1.50	4.13
Europe	87	35,722	7.82	3.83	11.65
MENA	19	14,292	2.79	0.34	3.13
S. America	32	8,513	3.01	0.90	3.91
N. America	26	48,199	4.79	5.86	10.65

[a]"MENA" is the Middle East and North Africa, "N." is North and "S. is South.
[b]Income per capita is annual gross domestic product per capita for 2005.
[c]Increasing and decreasing block structures result in different charges for different use levels. The 2010 GWI data lists average charges for a use level of 15 cubic meters or 3,963 gallons per month.

Table 2. Level of Municipal Water System Tariffs

Water and wastewater rates vary noticeably over the listed regions. Water rates are highest in Europe and North America and lowest in Africa and Asia. The average water rate in Europe is more than three times the water rate in Africa. Wastewater rates are highest again in Europe and North America and lowest in Africa and MENA. Combined rates are less than average in Africa, MENA, South America and Asia.

A standardized cost index allows a comparison of water rates relative to the revenue needed to cover operating, maintenance and capital costs (Komives et al., 2005). The index divides rates into the four categories shown in Table 3: insufficient or sufficient to cover operating and maintenance costs (O&M), sufficient to cover operating, maintenance and capital costs (O&M&C) and sufficient to cover costs in addition to minimum operating, maintenance and capital costs.

Costs vary depending on local and regional differences in wages and other prices, so the index sets different rates for less and more developed countries. The index does not include a fourth category of "Sufficient for Additional Costs" for more developed countries, so this threshold was set at $9.00 in these countries, double the rate needed to cover standard operating, maintenance and capital costs. The analysis applies the four less developed country cost categories to systems where mean income per capita is less than $10,000 per year in the 2010 GWI survey. It applies the more developed cost categories to systems with income per capita more than $10,000 per year.

Table 3 categorizes tariffs for the 121 systems in lower income regions, the 155 systems in higher income regions and all systems. Almost one-third of the tariffs in low-income regions and 8% of the tariffs in higher income areas are insufficient to cover only operating and maintenance costs. Fifty-seven percent of tariffs in developing countries are insufficient to cover the additional costs of capital. The data indicate that over all systems, only 14% recover revenue sufficient to cover more than standard operating, maintenance and capital costs with their current tariffs. This means that as many as 86% of the systems provide inadequate incentives for water conservation by failing to include non-financial opportunity costs.

Context	Insufficient for O & M [a]	Sufficient for O & M [a]	Sufficient for O & M & C [a]	Sufficient for Additional Costs
Less Developed	Less than $0.9	$0.9 to $1.8	$1.8 to $4.5	Greater than $4.5
More Developed	Less than $1.8	$1.8 to $4.5	$4.5 to $9.0	Greater than $9.0
Global Water Systems:				
Income < $10,000 (%)	31	26	36	7
Income > $10,000 (%)	8	28	45	19
All (%)	18	27	41	14

[a]"O & M" is operating and maintenance cost and "O & M & C" is operating, maintenance and capital cost.

Table 3. Cost Sufficiency of Municipal Water Rates

Figures 2 and 3 show that there is considerable variation in the adequacy of tariff levels within regions and countries as well. Much of the MENA is arid and the water opportunity costs are likely to be high. Figure 2 indicates that tariffs in 7 MENA systems are inadequate to cover operating and maintenance expenses, let alone encourage water conservation consistent with both financial and non-financial opportunity costs. Eleven MENA tariffs appear to cover financial costs to some degree. Six tariffs exceed standard financial costs. The rates in Jerusalem, Tel Aviv and Dubai appear high enough to include some portion of opportunity costs in addition to the immediate financial requirements of operation, maintenance and capital costs.

Figure 3 shows that tariffs in the United States tend to recover operating and maintenance costs, but at least 11 of the 19 systems shown set tariffs that are insufficient to cover capital costs. Four low tariff systems—Dallas, Las Vegas, Denver and San Antonio--are in arid regions where the opportunity cost of water is high, yet their tariffs fail to match the standard index for normal financial costs. Six of the 19 systems set rates adequate for revenues in excess of standard financial costs. The tariff for one city, San Diego—also in an arid region—exceeds the $9 level where tariff revenue may include a portion of non-financial water opportunity costs.

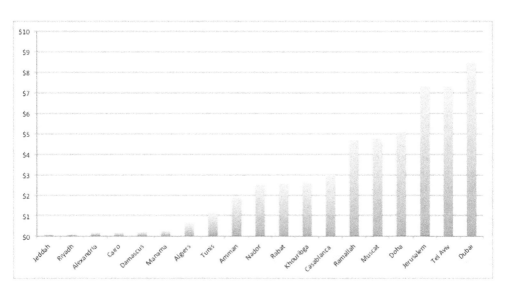

Fig. 2. Municipal Water Rates, Middle East and North Africa ($ per 1000 gallons)

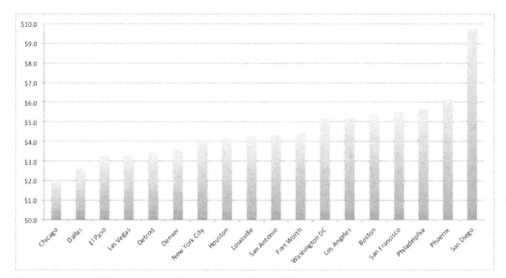

Fig. 3. Municipal Water Rates, United States ($ per 1000 gallons)

5. Implementing efficient residential tariffs

In many cities, large increases in water tariffs are likely to be required in order to encourage efficient water conservation. The amount of an increase in a particular water system depends on the current volumetric charge and the demand elasticities of water users. Dalhuisen et al. (2003) reports an average water demand elasticity of -0.4 in a review of 314 elasticity estimates obtained in 64 different research studies. However, elasticities varied significantly across studies and cities, so water use in a particular city may be less or more responsive to price increases than indicated by the average elasticity (Dalhuisen et al., 2003). For example, the average elasticity of -.4 means that reducing water use by 10% requires a 25% increase in a tariff volumetric charge. With a demand elasticity of -.1, reducing water use by 10% requires a 100% increase in a volumetric charge.

Water users are likely to resist large and unexplained tariff increases. Client acceptance of efficient water tariffs requires explanation and public education regarding the real economic costs of water. In some cases, there may be advantages to implementing efficient tariffs on a delayed schedule in order to give water users time to adjust and adopt water-saving habits and technologies before sustaining higher prices.

Three approaches to tariff reduce the financial impact of efficient water conservation incentives. The first approach is to use the efficient fixed charge to redistribute revenues in excess of financial costs. Tariffs based on unpaid, non-financial opportunity cost are certain to bring in surplus revenue above the revenue required to meet financial costs. Surplus revenue may be redistributed to water users through the fixed charge portion of the efficient tariff in a way that is consistent with fairness and equity concerns. As long as fixed charge rebates are not correlated with users' volumetric payment, fixed charge rebates do not distort the tariff incentive for efficient water conservation.

A second approach to increasing public acceptance of efficient water conservation incentives is to implement efficient tariffs for higher levels of water use and an inefficiently low tariff with volumetric water conservation rebate for low volume users (Collinge, 1994). Water conservation rebates for low volume users communicate the efficient incentive for water use *without* changes in volumetric charges below a certain threshold of water use. The volumetric rebate need only be set to pay users the volumetric opportunity cost for reduced water use. The threshold that distinguishes low and high water use may be set for individual users based on some percentage of historical use or it may be set at the same level for all water users based on some other criteria, such as using all surplus revenue encourage water conservation.

A two-tariff program combined conservation rebates may be structured in the following way. First, the volumetric charge is raised to an efficient level, λ^*, for water use in excess of the selected threshold for high water use. Second, for water use less than the threshold, the volumetric charge is set to λ^0, an amount less than λ^*. The lesser charge, λ^0, may leave unchanged an existing volumetric charge or it may be adjusted to some other level higher or lower than an existing volumetric charge. The only requirement is that $\lambda^0 \le \lambda^*$. The third step is to set a volumetric conservation rebate. The conservation rebate is equal to the difference between the efficient volumetric charge and the lower volumetric charge, $\lambda^* - \lambda^0$.

With the described charges and rebate in place, all water users face an efficient incentive for water conservation. Water users above the threshold pay the full opportunity cost, λ^*, on each unit of water used. When water users below the threshold use an additional unit of water, they give up the opportunity to earn the rebate, $\lambda^* - \lambda^0$, on that unit of water and they pay the volumetric charge, λ^0. The net payment for an additional unit of water use below the threshold is composed of two parts, (i) the sacrifice of the rebate, $\lambda^* - \lambda^0$, and (ii) the payment of the volumetric charge, λ^0. The sum of the sacrifice and volumetric charge is the opportunity cost of water use, the efficient incentive $\lambda^* = (\lambda^* - \lambda^0) + \lambda^0$.

The two-tariff-and-rebate program gives all water users an efficient incentive for water conservation. Water users below the threshold have an additional incentive to accept the program since they may pay the current volumetric charge and have an opportunity to earn rebate income. Setting thresholds at the individual level based on historical water use gives *all* users the prospect of earning rebates. Hence, it may be possible to make almost all users better off by including unpaid opportunity costs in an efficient water conservation tariff program.

A third approach requires no change in tariffs below a selected threshold of water use, but does require a physical infrastructure to transfer water between the residential water network and the next best alternative use. This is possible in regions such as the Rio Grande basin where both urban and agricultural water is stored and withdrawn from the same network of canals and storage reservoirs.

The third approach is based on resale of water between residential water users and agricultural water users (Haddad, 2000). Similar to the tariff and rebate program, a water use threshold is set either for residential users as a group or for each user based on historical water use. In addition, a water savings account is set up for each user within the water networks billing and accounting system.

Water users make deposits into the savings account by reducing their *planned* water use below the threshold selected by the water authority. The amount deposited is the difference between the selected threshold and the amount of water that an individual plans to use. Deposits can be made by telephone, by mail-in card or on the internet. The municipal system then acts as an agent for individual savers and sells the aggregate amount of planned savings to agricultural users or in-situ agents. Sales revenues are returned to savers in proportion to their *actual* savings.

Residential users pay the full opportunity cost of water in two situations. The first is when their water use exceeds the threshold selected by the water authority. The second, is when actual water saving is less than planned saving. Users that save less than they planned, pay the full opportunity cost of water on the difference between planned and actual savings. Apart from these two cases, users pay a volumetric rate that the water authority sets for water use that is below the threshold minus planned savings. Tariff charges for water use less than the threshold minus planned savings may remain at current inefficient rates.

The third approach communicates efficient water conservation incentives to all residential water users. Users using more than the threshold amount of water face the full opportunity cost for an additional unit of water. Users below the threshold amount forego the opportunity to sell water at its opportunity cost to agricultural users when they fail to 'save' water. Each type of user has an incentive to invest in water conservation consistent with its value in the next best use.

Like the two-tariff-and-rebate program, water savings accounts and resale give residential water users an opportunity to earn income from the true economic value of water. By setting appropriate thresholds at the level of an individual user, the income earning potential can be extended to all waters users. By saving water, users have the opportunity to converts financial pain into financial gain. Users thereby share the benefit of efficient water conservation.

6. Conclusions

Economic principles help identify the consequences of wasting scarce water. The opportunity cost concept shows wasting water is not just misguided. Wasting scarce water destroys real economic opportunities and leads to losses for other water users. The concept of water demand guides the measurement of water values, helps evaluate users' responses to conservation policies, and shows how to measure the deadweight loss of inefficient water use as well as the benefits of water conservation. Analysis of water trading shows how trade gives water users strong incentives to find the highest value uses of water and to eliminate waste by moving water into those uses. Third-party effects remind us that water use has potential consequences elsewhere in the hydrological cycle, such as on *in-situ* and downstream users.

The analysis used these economic principles to evaluate whether municipal water tariffs may be designed to encourage efficient water conservation. The analysis found that efficient water conservation tariffs have two parts, a volumetric charge that communicates the opportunity cost of an additional unit of water and a fixed charge that is adjusted at the user level to address equity and fairness and at the aggregate level to address revenue requirements not covered by the volumetric charge.

Empirical analysis showed that municipal water systems across the globe are large reservoirs of wasted water. More than 80% of 276 large water systems worldwide use tariffs that encourage water waste. There is some evidence of tariff reform in Europe, but, even there, the majority of water systems use inefficient tariffs that encourage wasted water. More than 85% of the 276 systems set tariffs so low that they appear unlikely to recover capital costs. Forty-five percent of tariffs fail to cover likely operating and maintenance costs. Systems in low-income cities appear most likely to use tariffs that disadvantage the poor, threaten financial viability and waste scarce and highly valued water.

Overall, the tariffs used by cities around world suggest rather solemn prospects for many water systems: possible financial insolvency, reduced service quality and service areas, and abundant water waste. Efficient water conservation tariffs can contribute to financial solvency and unlock the reservoirs of wasted water. Water is scarce and highly valued. An efficient water tariff communicates the high value of water.

Water users are likely to resist such communication in the form of unexplained increases in their water bills. Explanation and education campaigns are standard approaches to achieving clients' acceptance of water conservation policies. There are also two ways to modify tariff programs so that users can share the gains obtained from efficient water conservation. The first approach is a two-tariff program with a water conservation rebate. This two-tariff program sets a high tariff charge for water use beyond a certain threshold and a low tariff for water use below the threshold. The threshold may be set for an individual user based on historical use or it may be a single threshold for all users in the system. The volumetric charge in the high tariff is equal to the marginal opportunity cost of water. The low tariff can be set at any lower volumetric charge, including being left unchanged from an existing rate.

The conservation rebate equals the marginal opportunity cost of water. The rebate is paid to users based on water savings relative to the threshold — on the positive difference between threshold and the water a user actually uses. The two-tariff-and-rebate program presents all users with the full economic cost of using an additional unit of water. In contrast to an across-the-board tariff increase, users actually pay the full cost of water only on water use above the threshold. Water use below the threshold has a lower out-of-pocket cost, but the same opportunity cost. A user that fails to conserve below the threshold gives up both the rebate and volumetric charge on each additional unit of water use. Users above and below the threshold have a full and efficient incentive to invest in water conservation.

The second approach is similar to the two-tariff-and-rebate program. Instead of a rebate, the second approach offers users water conservation savings accounts. The savings accounts is an electronic entry maintained by the water systems billing and account system. Users make savings deposits by cutting back on water use. As savings accumulate, the water system sells saved water at its full opportunity cost to users outside the system, such as agricultural irrigators or trustees for environmental interests. Sales revenues are returned to savers in proportion to their water deposits. As in the two-tariff-and-rebate program, all water users face the full opportunity cost of an additional unit of water and each has an opportunity to earn income from water conservation.

7. References

Allan, A. (2003). A comparison between the water law reforms in South Africa and Scotland: Can a generic natural water law model be developed from these examples?. *Natural Resources Journal*, Vol. 43, No. 2, (Spring 2003) pp. 419-490, ISSN 0028-0739

Beecher, J.A., Bell, H., Bill, C., Brydon, B., Dief, M.K., Foley, T.D., Lowdon, A., Millan, A.M., & Walton, B. (2005). *Utility Rate Structures: Investigating International Principles and Customer Views*. Awwa Research Foundation, Denver, CO

Atwood, C., Kreutzwiser, R., & de Loe, R. (2007). Residents' assessment of an urban outdoor water conservation program in Guelph, Ontario. *Journal of the American*

Water Resources Association. Vol. 43, No. 2, (April 2007) pp. 427-439, ISSN 1093-474X

Banerjee, S., Foster, V., Ying, Y., Skilling, H., & Wodon, Q. (2008). *Cost Recovery, Equity, and Efficiency in Water Tariffs: Evidence from African Utilities*, The World Bank, Policy Research Working Paper 5384, Washington, DC

Barberan, R., & Arbues, F. (2009). Equity in domestic water rates design. *Water Resource Management*, Vol. 23, No. 10, (2009), pp. 2101-2118, ISSN 0920-4741

Beecher, J.A., & Kalmback, J.A. (2011). *2010 Great Lakes Water Survey*. Institute of Public Utilities at Michigan State University

Brewer, J. Glennon, R. Ker, A., & Libecap, G. (2008). 2006 presidential address water markets in the west: Prices, trading and contractual forms. *Economic Inquiry*, Vol. 46, No. 2, (April 2008), pp. 91-111, ISSN 0095-2583

Brookshire, D.S., Burness, H.S., Chermak, J.M., & Krause, K. (2002). Western Urban Water Demand. *Natural Resources Journal*, Vol. 42, No. 4, (Fall 2002), pp. 879-898, ISSN 0028-0739

Coase, R.H. (1946). The marginal cost controversy. *Economica*, Vol. 13, No. 51, (August 1942), pp. 169-182, ISSN 0013-0427

Collinge, R.A. (1994). Transferable Rate Entitlements: The overlooked opportunity in municipal water pricing. *Public Finance Quarterly*, Vol. 22, No. 1, (January 2004), pp. 46-64, ISSN 1091-1421

Coman, K. (2011). Some unsettled problems of irrigation. *American Economic Review*, Vol. 1, No. 1, (March 1911), pp. 1-19, ISSN 0002-8282

Crane, R. (1994). Water markets, market reform and the urban poor: Results from Jakarta, Idonesia. *World Development*, Vol. 22, No. 1, (January 1994), pp. 71-83, ISSN 0305-750X

Dalhuisen, J.M., Florax, R.J.G.M., de Groot, H.L.F.M., & Nijkamp, P. (2003). Price and income elasticities of residential water demand: A meta-analysis. *Land Economics*, Vol. 79, No. 2, (May 2003), pp. 292-308, ISSN 0023-7639

De Mouche, L., Landfair, S. & Ward, F.A. (2011). Water right prices in the Rio Grande: Analysis and policy implications. *International Journal of Water Resources Development*, Vol. 27, No. 2, (June 2011), pp. 291-314, ISSN 0790-0627

Dziegielewski, B., Kiefer, J., & Bik, T. (2004). *Water Rates and Ratemaking Practices in Community Water Systems in Illinois*, Southern Illinois University, ISBN, Carbondale, IL

Gaudin, S. (2006) Effect of price information on residential water demand. *Applied Economics*, Vol. 38, No. 4, (March 2006), pp. 282-393, ISSN 0003-6846

Grafton, Q.R., Landry, C., Libecap, G.D., McGlennon, S., & O'Brien, R. (2010). An Integrated Assessment of Water Markets: Australia, Chile, China, South Africa and the USA, NBER Working Paper No. 16203, National Bureau of Economic Research, Cambridge, MA

Griffin, R.C. (2001). Effective water pricing. *Journal of the American Water Resources Association*. Vol. 37, No. 5, (October 2001), pp. 1335-1347, ISSN 1752-1688

Griffin, R.C., & Boadu, F.O. (1992). Water marketing in Texas: Opportunities for reform. *Natural Resources Journal*, Vol. 32, No. 2, (Spring 1992), pp. 265-288, ISSN 0028-0739

Haddad, B. (2000). Economic incentives for water conservation on the Monterey Peninsula: The market proposal. *Journal of the American Water Resources Association*, Vol. 36, No. 1, (February 2000), pp. 1-15, ISSN 1093-474X

Hall, D.C. (2009). Politically feasible, revenue sufficient, and economically efficient municipal water rates. *Contemporary Economic Policy*, Vol. 27, No. 4, (October 2009), pp. 539-554, ISSN 1465-7287

Hickey, J.T., & Diaz, G.E. (1999). From flow to fish to dollars: An integrated approach to water allocation. *Journal of the American Water Resources Association*, Vol. 35, No. 5, (October 1999), pp. 1053-1067, ISSN 1093-474X

Hoehn, J.P., & Krieger, D.J. (2000). Economic analysis of water service investments and tariffs in Cairo, Egypt. *Evaluation Review*, Vol. 126, No. 6, (November-December 2000), pp. 345-350, ISSN 0733-9496

Ipe, V.C., & Bhagwat, S.B. (2002). Chicago's water market: dynamics of demand, prices and scarcity rents. *Applied Economics*, Vol. 34, No. 17, (November 2002), pp. 2157-2163, ISSN 0003-6846

Kallis, G. (2008). Droughts. *Annual Reviews of Environment and Resources*, Vol. 33, pp. 85-118, ISSN 1543-5938

Komives, K., Foster, V., Halpern, J., Wodon, Q., & Abdullah, R. (2005). *Water, Electricity and the Poor: Who Benefits from Utility Subsidies?*, The World Bank, ISBN 9780821363423, Washington, DC.

Kenney, D.S., Goemans, C., Klein, R., Lowrey, J., & Reidy, K. (2008). Residential water demand management: Lessons from Aurora, Colorado. *Journal of the American Water Resources Association*, Vol. 44, No. 1 (February 2008), pp. 192-207, ISSN 1752-1688

Libecap, G.D. (2007). *Owens Valley Revisited*, Stanford University Press, ISBN 9780804753791, Stanford, CA

Martins, R., & Fortunato, A. (2007). Residential water demand under block rates – a Portuguese case study. *Water Policy*, Vol. 9, No. 2, (2007), pp. 217-230, ISSN 1366-7017

Nataraj, S., & Hanemann, W.M. (2011). Does marginal price matter? A regression discontinuity approach to estimating water demand. *The Journal of Environmental Economics and Management*, Vol. 61, No. 2, (March 2011), pp. 198-212, ISSN 0095-0696

Nauges, C., & Whittington, D. (2009). Estimation of Water Demand in Developing Countries: An Overview. *The World Bank Research Observer*, Vol. 25, No. 2, (November 2009), pp. 263-294, ISSN 0257-3032

Olmstead, S.M. (2010). The economics of managing scarce water resources. *Review of Environmental Economics and Policy*, Vol. 4, No. 2, (Summer 2010), pp. 179-198, ISSN 1750-6816

Olmstead, S.M., Hanemann, W.M., & Stavins, R.N. (2007). Water demand under alternative price structures. *Journal of Environmental Economics and Management*, Vol. 54, No. 2, (September 2008), pp. 181-198, ISSN 0095-0696

Olmstead, S.M. & Stavins, R.N. (2009). Comparing price and nonprice approaches to urban water conservation. *Water Resources Research*, Vol. 45, No. 4, (April 2009), ISSN 0043-1397

Organization for Economic Co-Operation and Development. (2009). *Managing Water for All: An OECD Perspective on Pricing and Financing*, OECD Publications, ISBN 9789264050334

Ostrom, E. (1990). *Governing the Commons: The Evolution of Institutions for Collective Action.* Cambridge University Press, ISBN 0521405998

Platt, J. (2001). *Economic Nonmarket Valuation of Instream Flows*. United States Department of Interior, Bureau of Reclamation, Technical Memorandum Number EC-2001-01, Denver, CO

Renwick, M.E., & Green, R.D. (1999). Do residential water demand side management policies measure up? An analysis of eight California water agencies. *Journal of Environmental Economics and Management*, Vol. 40, No. 1, (July, 2000), pp. 37-55, ISSN 0095-0696

Rogerson, C.M. (1996). Willingness to pay for water: The international debates. *Water SA*, Vol. 22, No. 4, (October 1996), pp. 373-380, ISSN 0378-4738

Ruml, C.C. (2005). The Coase Theorem and Western U.S. appropriative water rights. Natural Resources Journal, Vol. 45, No. 1, (February 2005), pp. 169-200, ISSN 0028-0739

Saleth, M.R., & Dinar, A. (2001). Preconditions for market solution to urban water scarcity: Empirical results from Hyderabad City, India. *Water Resources Research*, Vol. 37, No. 1, (January 2001), pp. 119-131, ISSN 0043-1397

Slaughter, R.A. (2009). A transactions cost approach to the theoretical foundations of water markets. *Journal of the American Water Resources Association*, Vol. 45, No. 2, (April 2009), pp. 331-342, ISSN 1752-1688

Snellen, W.B., & Schrevel, A. (2004). IWRM: for sustainable use of water 50 years of international experience with the concept of integrated water management, Ministry of Agriculture, Nature and Food Quality, The Netherlands, Wageningen, Netherlands

State of California. (2008). *Urban Drought Guidebook 2008 Updated Edition*, State of California, Sacramento, CA

Sunding, D., & Chong, H. (2006) Water markets and trading. *Annual Review of Environment and Resources*, Vol. 31, (November 2006), pp. 239-264, ISSN 1543-5938

Spangler, D.K. (2007). Utah farmers are told to use water or lose it, *Deseret News* (May 27 2004), Salt Lake City, UT, retrieved from www.deseretnews.com/article/595066072/Utah-farmers-are-told-to-use-water-or-lose-it.html

Western Resource Advocates, Colorado Environmental Coalition, & Western Colorado Congress. (2004). *Water Rate Structures in Colorado: How Colorado Cities Compare in Using this Important Water Use Efficiency Tool*, Western Resource Advocates, Boulder, CO

Whitcomb, J. (2005). *Florida Water Rates Evaluation of Single-Family Homes*, prepared for the Southwest Florida Water Management District, Brooksville, FL, St. Johns River Water Management District, Palatka, FL, South Florida Water Management District, West Palm Beach, FL, and the Northwest Florida Water Management District, Havana, FL

World Commission on Dams. (2000). *Dams and Development: A New Framework for Decision Making*, Earthscan Publications Ltd, ISBN 1853837970, Sterling, VA

World Health Organization & United Nations Children's Fund. (2000). *Global Water Supply and Sanitation Assessment 2000 Report*, World Health Organization, United Nations Chidren's Fund, ISBN 9241562021, (New York)

Worthington, A.C., & Hoffman, M. (2008). An empirical survey of residential water demand modeling. *Journal of Economic Surveys*, Vol. 22, No. 5, (December 2008), pp. 842-871, ISSN 0950-0804

7

Water Management in the Petroleum Refining Industry

Petia Mijaylova Nacheva
Mexican Institute of Water Technology
Mexico

1. Introduction

Petroleum refining industry uses large volumes of water. The water demand is up to 3 m³ of water for every ton of petroleum processed (US EPA, 1980, 1982; WB, 1998). Almost 56% of this quantity is used in cooling systems, 16% in boiling systems, 19% in production processes and the rest in auxiliary operations. The water usage in the Mexican refineries is almost 155 millions m³ per year; it is 2.46 m³ of water per ton of processed petroleum (PEMEX, 2007). The water supply and distribution for the different uses depend on the oil transformation processes in the refineries, which are based on the type of crude petroleum that each refinery processes and on the generated products. The cooling waters are generally recycled, but the losses by evaporation are high, up to 50% of the amount of the used water. The reduction of the losses and the increase of the cycles of recirculation represent an area of opportunities to diminish the water demand. The requirements with respect to the quality of the water used in the cooling systems are not very strict (Nalco, 1995; US EPA, 1980), which makes possible to use treated wastewater as alternative water source (Sastry & Sundaramoorthy, 1996; Levin & Asano, 2002). The water for the production processes and for services must be of high quality, equivalent to the one of the drinking water. For the boilers and some production processes, the water must be in addition demineralized (Powel, 1988; Nalco, 1995). The Mexican refineries have demineralizing plants which generally use filtration and ion exchange or reverse osmosis systems.

The quantity of the wastewater generated in the refineries is almost 50% of the used fresh water (US EPA, 1982; WB, 1998; EC, 2000). Different collection systems are used in the refineries, depending on the effluent composition and the point of generation. The waters that are been in contact with petroleum and its derivatives contain oil, hydrocarbons, phenols, sulfides, ammonia and large quantities of inorganic salts (US EPA, 1995; Mukherjee et al., 2011). Following the implemented production processes, organic acids, dissolving substances and aromatic compounds may by also present in the wastewater. These effluents are conducted by means of an oily drainage towards the pre-treatment systems for the oil and oily solids separation. The optimization of the production processes, the appropriate control of the operation procedures and the implementation of appropriate water management practices have yield significant reductions of the wastewater flows and of the level of the contaminant loads. Consequently the quality of wastewater discharges can be improved reducing this way their environmental impact and the treatment costs (IPIECA,

2010). Ones of the first recommendations were with regard to the management of sour water and spent caustics (US EPA, 1982, 1995; WB, 1998; EC, 2000). The sour waters that contain ammoniac, phenol, hydrogen sulphide and cyanides require previous treatment before being mixed with other effluents. Spent caustics that contain sulfides, mercaptans and hydrocarbons must be also collected and treated individually.

The waters that do not have been in contact with petroleum are collected by means of separated drainages (EC, 2000). This is the case of the cooling towers blowdowns that basically contain dissolved or suspended mineral salts, as well as the effluents from filter backwashings and resin regenerations or the inverse osmosis rejections. The concentrates discharges from the resin regeneration and the inverse osmosis rejections require a special management, whereas the cooling towers blowdowns and the effluents from filter backwashings need only a slight treatment and after this they can be successfully reused (US EPA, 1982). The sanitary wastewaters are also treated individually. Surface water runoff is generated in the refineries during the raining periods. Special sewage system is constructed for the recollection and conduction of this water. Theoretically this sewage system does not receive contaminated waters, nevertheless some accidental spills and discharges can be received. That is why retention tanks are constructed for these waters to remove the main pollutions, oil and solids.

The oily wastewater is the most contaminated effluent of the above described. After the pretreatment, the wastewaters must be submitted to biological and advanced treatments for accomplishment of the requirements for discharge in the receiving body (WB, 1998; Eckenfelder, 2000; EC, 2000). The effluent obtained after the advanced treatment is apt for reuse in the cooling system, compensating therefore the losses by evaporation. It may be also used in other processes and services of the refinery. This way, besides reducing the water consumption, the danger of contamination of the receiving bodies can be eliminated. The first pretreatment process of the oily wastewater is the oil-water separation. The conventional rectangular-channel separators, developed by the American Petroleum Institute (API) are wildly used for this purpose, and their design criteria are summarized in the publication API, 1990. Many other separators had been developed based on the oil-water separation theory and some of them, as the parallel plate and corrugated plate separators, had been implemented in the petroleum refineries (WEF, 1994). The oil separators remove only the fraction of free oil; the emulsified and the dissolved oil remain in the separator effluent. Therefore, destabilization of oil-water emulsions followed by separation by dissolved air flotation (DAF) is required for the further pretreatment of the oily wastewaters (Eckenfelder, 2000; Galil & Wolf, 2001; Al-Shamrani et al., 2002). Different biological treatment processes have been used for refinery wastewater treatment, such as aerated ponds, activated sludge, biological contactors, sequential bath reactors and moving bed reactors (Galil & Rebhun, 1992; Baron et al., 2000; Lee et al., 2004; Schneider et al., 2011). The first researches that had been done for recycling of the biologically treated refinery effluent involved: activated carbon adsorption alone or in combination with ozonation or sand filtration (Miskovic et al., 1986; Guarino et al., 1988; Farooq & Misbahuddin, 1991). The membrane technology development allowed additional options, such as ultrafiltration and reverse osmosis (Zubarev et al., 1990; Elmaleh & Ghaffor,1996; Teodosiu et al.,1999; Daxin Wang et al., 2011). The implementation of the advanced treatment technology allowed reusing of the biologically treated wastewater and freshwater savings in the refineries. Baron et al. (2000) reported a case study of water management project for the use of reclaimed wastewater in one Mexican refinery. Lime softening and filtration were implemented for the advanced treatment of the secondary effluent. The use of seawater as alternative fresh water source was considered in this project. Reverse osmosis (RO) system was installed for the seawater demineralization and the performed evaluation indicated that the RO facility assures the Refinery a reliable water supply resulting in reduction of the freshwater consume.

The objective of the presented here study was to develop appropriate water resource management options for reaching complete wastewater reuse and water use minimization in two Mexican refineries. The technological feasibility of the wastewater reuse was based on evaluation of the current wastewater treatment performance and experimental tests on alternative treatment processes with a view to improve the quality of the reclaimed water and enable its recycling.

2. Methodology

The study of the refinery wastewater treatment for reuse began with the characterization of the main effluents. Evaluation of the current wastewater treatment systems were performed based on three samplings performed in different periods of the year. The following parameters were considered: Oil and Grease (O&G), Chemical Oxigen Demand (COD), Soluble Chemical Oxigen Demand ($COD_{soluble}$), Biochemical Oxigen Demand (BOD_5), Total Suspended Solids (TSS), Total Dissolved Solids (TDS), Phenols, Ammonium Nitrogen (NH_4-N), Total Kjeldahl Nitrogen (TKN), Total Phosphorus (P_{total}), S^{2-}, Hardness, Alkalinity, pH, Conductivity, SO_4^{2-}, F^-, Cl^-. Based on the obtained characterizations, appropriate water handling options were analyzed. Treatability tests were performed for all of the proposed treatment processes to obtain the values of the design parameters. The performance of gravity oil-water separators varies with changes in the characteristics of the oil and wastewater, including flow rate, specific gravity, salinity, temperature, viscosity, and oil-globule seize (API, 1990). That is why tests for natural flotation were performed in situ using acrylic columns with 0.25 m diameter and 2.5 m high. Sampling taps were located at 0.5 m depth intervals. The columns were felt with the tested wastewater and samples were drawn off at selected time intervals up to 120 min. The samples were analyzed for O&G and TSS. Additional samples for COD were obtained for the study in refinery R2. The results were expressed in terms of percent removal at each tap and time interval. These removals were plotted against their respective depth and times and the flotation and settling curves were obtained. Then the data were used to develop the removal-surface loading rate relationships.

The destabilization of oil-water emulsions was studied by means of jar tests in an equipment Philips y Bird PB 700. Different mineral coagulants, polymers and their combinations were evaluated in the effluents from the oil separators. The commercialized products were: Aluminium sulphate (SAS), polyaluminium chloride (PAX-XL19, PAX-260XLS, PAX-16S, PAX-XL60S), ferric chloride (PIX-111), ferric sulphate (PIX-145 and Ferrix-3). The coagulants were tested individually and combined with polymers. The following anionic polymers were used: OPTOFLOC A-1638 and AE-1488 (high molecular weight and high charge density); SUPERFLOC A-100 HMW (high molecular weight and moderate charge density) and PHENOLPOL A-305 (high molecular weight and low charge density). Cationic polymers were: SUPEFLOC C-1288, C-1392, C-1781 and LACKFLOC-C-5100 (high molecular weight and high charge density); SUPERFLOC C-498 (moderate molecular weight and high charge density); ECOFLOC (high molecular weight and moderate charge density). The test conditions during the study in refinery R1 were: rapid mixing at 120 rpm during 3 min, slow mixing at 30 rpm during 20 min, separation time of 25 min. The tests with the effluents in refinery R2 were performed as follows: rapid mixing at 150 rpm during 3 min, slow mixing at 20 rpm during 15 min, separation during 30 min. The effect of wastewater acidification and alcalinization was first determined using H_2SO_4 and NaOH. Then tests with dose variation were carried out and the best product and dose were selected for each case. The pH effect on the removal efficiency was determined for some of the tested products. The analyzed parameters were O&G, COD and TSS. Turbidity and color were also followed in the refinery R2 study.

Once selected the best chemical reagents, the separation process of the formed flocks and oil with dissolved air flotation (DAF) was evaluated. A bench scale DAF unit consisting of an compressor, a 3 L stainless steel unpacked saturator vessel and a 5 L flotation cell was used. The flotation cell has a variable speed-controlled impeller providing rotational speeds between 100-300 and 20-100 rpm for rapid and slow mixing respectively. The process of dissolved air flotation was studied with previous flocculation. The tested wastewater was introduced to the flotation cell which was first used for the flocculation. The flocculant was added and mixed with the wastewater for 3 min at 150 rpm, followed by slow mixing for 15 min at 20 rpm for flocculation. At the end of the flocculation process the saturator vessel was connected to the flotation cell in order to transfer a controlled amount of previously pressurized treated water. At that moment the flotation was allowed to proceed. When released to the open cell, the dissolved air was transformed into a mass of fine air bubbles, which could attach to the flocs and carry them to the upper liquid surface. After determined retention time, samples of the treated water were collected for analysis. Two experimental runs were carried out with oily wastewater from refinery R1 and one with water from refinery R2. Chemical reagents, recycling ratio (R) and saturation pressure (P) were the variables during the first experimental run. Initial O&G concentration, P and R were the variables during the second run. Factorial experimental designs 2^3 were used in the first experimental run, adding central points for P and R. ANOVA was applied for the analysis of the obtained results. Experimental design 3^3 with two central points for P and R was used in the second run. The tests performed for refinery R2 used 3^3 experimental design and the variables were: P, R and HRT. The output parameters were O&G, COD, TSS, turbidity and color. All analytical procedures were based on the *Standard Methods for Examination of Water and Wastewater,* (2005). The biological and the advanced processes were evaluated based on the reports provided by the real scale facilities. The obtained water qualities of the effluents from the evaluated treatment processes were compared with the required ones for different kinds of reuse. Finally, the feasibility of the proposed water reuse options was determined for each refinery.

3. Results and discussion

3.1 Water consumption, wastewater characteristics and evaluation of the current pretreatment systems

Surface water, such as water from river, reservoir and lagoon, are the main water sources for both studied refineries (R1 and R2). The current water consumption and the fresh water distribution for the different uses are presented in Table 1. The wastewater quantities represent 48% of the consumption in both refineries. There are two main oily effluents in each refinery and both refineries have separate treatment of the sour waters and for the spent caustics. The refinery R1 has three stage oil separators. The discharge with the highest oil content passes through First Stage Separator (S1); the effluent from this separator is mixed with the second oily discharge and the mixture passes through the second (S2) and third stage separators (S3). The characteristics of the main oily effluents (D1 and D2) are presented in Table 2. The high O&G concentration in the oily wastewater indicates the necessity of prevention measures, such as process optimization and control implementation.

Refinery	Fresh water consumption		Water distribution per uses, %				
	Water-flow, L/s	Consumption, m³/t processed petroleum	Cooling tower make-up	Boiler make-up and power generation	Production processes	Service water
R1	384	2.10	58.1	19.5	11.9	10.5
R2	467	2.28	59.7	18.8	14.3	7.1

Table 1. Water consumption and uses in the studied refineries

Parameter	Oily discharge D1	Oily discharge D2	Efluent from S1	Efluent from S2	Efluent from S3
Flow, L/s	49±9	50±10	49±7	99±19	99±19
Temperature, °C	37±6	36±5	36±5	35±4	34±4
O&G, mg/L	11,455±5,230	7,880±4,870	2,291±1,350	69±7	27±5
COD, mg/L	8,316±2,980	6,806±1,990	2,245±1,105	1,390±228	448±81
TSS, mg/L	496±78	376±65	233±45	207±22	28±9
TDS, mg/L	964±248	1,390±295	894±196	1,160±220	1,138±206
Sulphates, mg/L	255±38	424±49	243±32	319±39	280±35
Chlorides, mg/L	249±47	119±22	229±34	230±37	228±33
Sulphides, mg/L	37±22	59±34	36±20	37±11	6±5
Fluorides, mg/L	3.5±2.2	4.3±3.2	3.5±2.4	5.3±2.2	2.6±2.1
Phenols, mg/L	0.40±0.44	1.63±0.85	0.37±0.25	0.51±0.32	0.22±0.21
NH₄-N, mg/L	7.0±6.5	15.3±3.4	6.9±5.1	12.4±5.5	12.3±6.6
TKN, mg/L	12.4±7.9	28.0±9.2	11.2±6.1	20.4±8.5	20.3±7.4
Alkalinity, mg/L	133±38	200±55	132±26	149±44	132±29
Hardness, mg CaCO₃/L	337±45	532±26	330±34	412±44	347±32
pH	7.20±0.12	7.22±0.11	7.15±0.13	7.33±0.11	7.38±0.10
Conductivity, µS/cm	2,570±387	1,989±266	2,375±306	2,250±278	2,153±255

Table 2. Characteristics of the oily effluents in refinery R1

The evaluation of the oil wastewater pretreatment indicated that the first stage separator provided average removals of 80%, 73% and 53% for O&G, COD and TSS respectively. The second stage separator present higher O&G removal, of 99%, the COD removal was of 69%, however the TSS removal was only 32%. The third stage separator has high hydraulic residence time, of 37 h and this contribute to an additional removal of O&G, COD and TSS of 61%, 68% and 86% respectively. Sulphides and phenols were partially removed in the separators. The rest of the components were not removed, precipitation phenomena were not observed. The oil specific gravity was determined of 0.92-0.95 (17-22°API) which allows the theoretic calculation of 0.07-0.11 cm/s rise rate of the oil globules with 0.15 mm diameter.

The three stage oil separators were well designed, considering all API recommendations (API, 1990); however the second and third stage separators are designed for flows 10 times higher than the real ones. The relatively low O&G removal obtained in the first stage separator is attributed to the deficient equipment for oil and sludge separation. The equipments of the second and third stage separators are also deficient and the obtained removals are attributed to the high retention capacity. Recommendation of better process control actions were made for the reduction of the oil concentrations in the wastewaters.

The refinery R1 has also two additional discharges. One of them (DS) is from a collector for mixture of sanitary discharges, cooling towers blowdowns and effluents from filter backwashings (average flowrate of 50 L/s). This wastewater has low COD and O&G, averages of 120 and 8 mg/L respectively; the TSS and TDS concentrations are 143 and 1,536 mg/L respectively. This effluent is currently discharged to the see without treatment; however TSS removal has to be implemented before its disposal.

The second additional discharge (D3) is from the area for crude petroleum storage and from oil demineralization (average flowrate of 13 L/s). This wastewater contains oil (980 ± 490 mg/L) and high salinity, which is attributed basically to the chlorides ($2,332\pm254$ mg/L). The effluent is submitted to a pretreatment in corrugated plate separator and after this is discharged to the see. It has to be mentioned that a lot of organic matter is still present in the effluent after the oil separation, average COD of 783 mg/L and phenols of 0.13 mg/L were determined. Thus, this effluent needs additional treatment before its final disposal.

The refinery R2 has two API separators, one for each oily wastewater discharge. Corrugated plate separators (CPS) are used as a second separation stage. The characteristics of the oily wastewaters and of the effluents from the separators are presented in Table 3. The O&G concentrations were significantly lower compared with the determined in the oily wastewaters generated in the refinery R1. The oil specific gravities were determined of 0.897 (24°API) and 0.951 (16°API) for discharge 1 and 2 respectively. The theoretic rise rates were calculated of 0.17 and 0.07 cm/s respectively, considering 0.15 mm oil globules and the minimal temperatures for each discharge. The fraction of soluble COD was 25-40% of the total COD. The high salinity of the oily discharge 1 is due to effluents from oil desalination processes. The salinity is attributed basically to the chlorides. The values of the BOD_5 were 24-15% of the COD. Ammonia nitrogen represented 52-57% of the TKN in the wastewater.

The performed evaluation indicated that the average O&G removals in both API separators were of 91%. The TSS removals were 87 and 78% in API 1 and API 2 respectively. The COD removals were 49 and 67% respectively. Hardness, TDS and chloride removals (22-36%) were observed in the API separator for discharge 1, which can be attributed to precipitation caused by the high water temperature. The sulphide removals in both API separators can be

Parameter	Oily discharge D1	Oily discharge D2	Efluent from API discharge 1	Efluent from API discharge 2	Efluent from the final CPS
Flow, L/s	116±55	108±72	113±46	107±69	220±109
Temperature, °C	44±7	32±2	41±4	32±1	38±3
O&G, mg/L	624±728	474±464	55±54	40±21	48±40
COD, mg/L	586±212	591±214	311±73	318±56	314±61
CODsoluble, mg/L	217±63	159±48	192±52	141±45	167±46
BOD$_5$, mg/L	144±54	146±84	108±26	102±60	105±52
TSS, mg/L	185±65	195±64	24±2	42±24	33±6
TDS, mg/L	1,583±250	828±167	1,076±155	733±109	883±165
Sulphates, mg/L	111±8	253±52	98±14	214±88	164±77
Chlorides, mg/L	782±39	241±86	545±83	222±89	388±85
Sulphides, mg/L	50±37	18±9	40±33	14±5	27±18
Fluorides, mg/L	0.50±0.08	0.36±0.14	0.39±0.14	0.35±0.16	0.37±0.11
Phenols, mg/L	0.95±0.65	1.29±0.82	0.82±0.61	1.21±0.90	1.01±0.72
NH$_4$-N, mg/L	28±22	35±32	25±21	33±36	29±23
TKN, mg/L	49±25	67±29	36±24	58±38	46±31
Ptotal, mg/L	0.70±0.17	0.87±0.22	0.63±0.16	0.72±0.13	0.66±0.15
Alkalinity, mg/L	123±21	102±30	105±38	100±25	104±40
Hardness, mg CaCO$_3$/L	389±126	224±35	249±38	207±45	225±34
pH	7.13±0.34	7.06±0.15	7.09±0.12	7.05±0.10	7.09±0.11
Conductivity, µS/cm	2,570±419	1,340±436	1,840±151	1,170±240	1,790±110

Table 3. Characteristics of the oily effluents in refinery R2

attributed basically to desorption. The evaluation indicated that the API separators were correctly designed; there was 40% additional capacity for safety reasons. However, the oil recollection and recovery, as well as the sludge extraction were deficient and reengineering project of the pretreatment facilities was developed, based on the wastewater characterizations and on the results of the performed treatability tests. The existing CPS did not provide any O&G, COD and TSS removal. The plate modules, after a complete cleaning, got saturated with oily sludge in few months. The constant cleaning and sludge extraction was too complicated operationally.

The obtained characterizations and the pretreatment performance evaluation indicated that additional treatment is required after the API separators for reaching the appropriate water quality for reuse. The emulsified and dissolved oil remain in the water after the physical separation. Therefore, as it had been indicated in previous publications (Eckenfelder, 2000; Galil & Wolf, 2001; Al-Shamrani et al., 2002), destabilization of the oil-water emulsions and separation by dissolved air flotation, followed by biological and advanced treatment are needed for an effective water reuse implementations.

3.2 Water management options

With the proposal to achieve a complete wastewater reuse and increase the fresh water saving in each one of the studied refineries, new water management options were suggested. The option development was based on the current water usage and management data, on the performed wastewater measurements and characterizations, as well as considering the results of the evaluation of the existing treatment systems.

The water management option for refinery R1 considered the treatment for reuse of the two effluents that are currently discharged to the sea. This refinery has already constructed sequential batch reactors, lime softening reactors, rapid sand filters and reverse osmosis system with a capacity of 86 L/s. These facilities require adjustment for the processing of all the pretreated wastewater. Currently only 50 L/s of the pretreated effluent are submitted to the biological treatment. The effluent is mixed with fresh water and then submitted to the advanced treatment. Performance problems in the separators frequently cause reductions of the influent to the biological treatment for avoiding biomass intoxication.

The current and the proposed new water management systems for the refinery R1 are presented on Fig. 1. Currently the refinery reuses only 30% of the generated wastewaters, which allowed 13% reduction of the fresh water consume. The proposed water management system considers complete reuse of the treated wastewater which will provide an increase of the fresh water save to 39%. Recently, a new municipal wastewater treatment facility was constructed next to the refinery with a capacity of 45 L/s. This facility included nitrification-denitrification activated sludge system with the objective to use the treated water in the cooling tower make-up in the refinery. This way 51% fresh water consume reduction will be reached.

The refinery R2 has already constructed nitrification-denitrification activated sludge system, followed by ultrafiltration and inverse osmosis systems. Currently this facility provides treatment to only 40-50% of the generated wastewater because of the high O&G concentrations in the effluent from the pretreatment system. The industrial effluent is mixed with 30 L/s domestic wastewater before to be submitted to the biological treatment. The obtained water use reduction was only 26%.

The current and the proposed new water management systems for the refinery R2 are presented on Fig. 2. The reengineering project for the pretreatment wastewater treatment system will provide a complete wastewater reuse and this way 59% fresh water consume reduction will be reached.

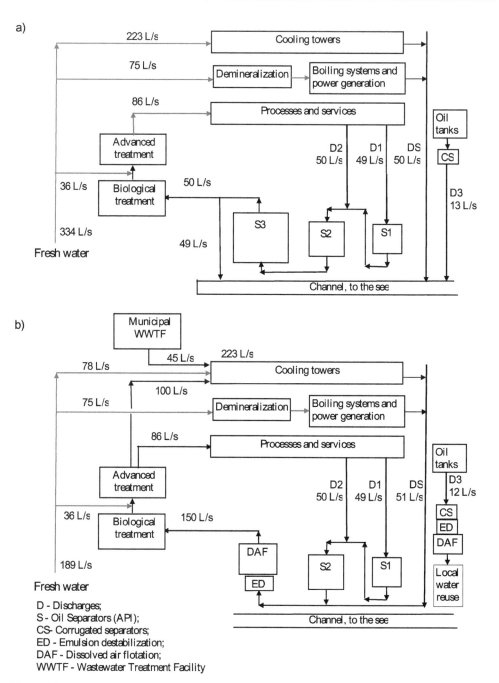

Fig. 1. Water management systems in the refinery R1: *a)* current management; *b)* proposed water management.

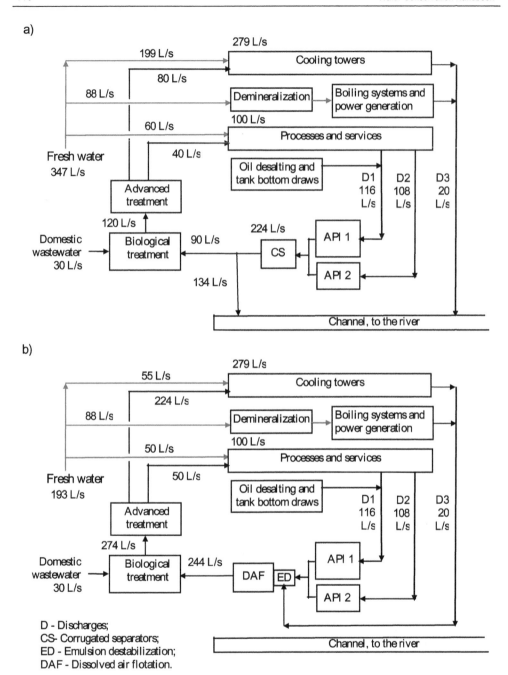

Fig. 2. Water management systems in the refinery R2: *a)* current management; *b)* proposed water management.

3.3 Results of the treatability tests

Treatability tests for natural oil flotation were performed in both refineries. For refinery R1 water samples for the tests were taken from the oily discharge 1 (influent to the first stage separator) and from the influent to the secondary stage separators which is a mixture of the oily discharge 2 with the effluent from the first stage separator. For refinery R2 water samples were taken from both oily discharges D1 and D2. The obtained removal-surface loading rate relationships for the refinery R1 are presented on Fig.3. As it can be seen, 90% O&G removal was obtained in the first and second stage separators with surface loading rates of 3.43 and 4.60 $m^3.m^{-2}.h^{-1}$ (floatation velocity of 0.10 and 0.13 cm/s) respectively. The simultaneous TSS removal was of 59% and 60% respectively with 30-40 min hydraulic retention time (HRT). Higher O&G removals, of 95% were obtained with surface loading rates of 1.15 and 1.53 $m^3.m^{-2}.h^{-1}$ (0.03 and 0.04 cm/s) respectively. The TSS removal did not increase substantially, 62% were obtained for both kinds of wastewater with HRT of 1.5-2.0 hours.

The results of the tests for natural oil flotation performed in refinery R2 are presented on Fig.4. O&G removals of 90% were obtained in D1 and D2 with surface loading rates of 2.77 and 2.30 $m^3.m^{-2}.h^{-1}$ (floatation velocity of 0.08 and 0.06 cm/s) respectively. The TSS removals were 68% and 59% respectively with 50-60 min HRT. The COD removals were relatively low, 34% and 32% respectively. O&G removals of 95% were obtained with the water of both discharges at surface loading rates of 1.15 $m^3.m^{-2}.h^{-1}$ (0.03 cm/s). The TSS and COD removals increased at this rate when the HRT of 2 h was used. TSS removals were 72% and 63% for D1 and D2 respectively; COD removals reached 39 and 34% respectively. The experimentally obtained floatation velocity was two times lower than the theoretically calculated for D1. Both velocities were similar in the case of D2. The tests indicated also that after the natural flotation the COD values remain in the range 340-460 mg/L, in spite of the low O&G concentrations (47-62 mg/L). The optimal separator depth was also obtained in the tests, it was 0.8-1.3 for the best O&G and COD removal and it could by up to 2.3 m considering as criteria the TSS removal.

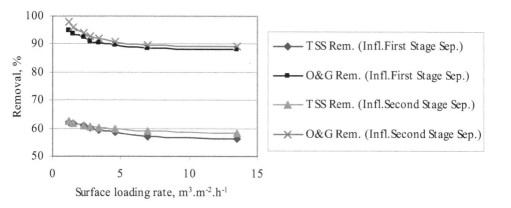

Fig. 3. Results of the treatability tests for natural flotation performed in Refinery R1.

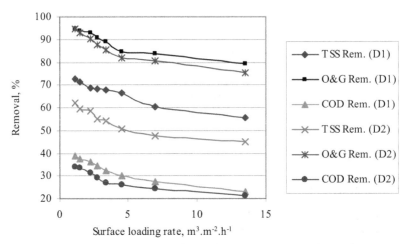

Fig. 4. Results of the treatability tests for natural flotation performed in Refinery R2.

The emulsion destabilization study began with preliminary tests applying only acidification and alcalinization of the wastewater. Fig. 5 shows the effect of the final pH on the O&G and COD removal in effluents from API separators. The average initial pH in the three effluents was 7.3±0.1. The effluent from the second stage separator of the refinery R2 had O&G and COD of 95 and 1,513 mg/L respectively. The effluents from the API separators of the refinery R2 had lower concentrations. The effluent API-D1 had O&G and COD of 58 and 518 mg/L respectively, the effluent API-D2 had 48 and 487 mg/L respectively. The results showed different comportment in the wastewater from refinery R1 and R2. The removals decreased gradually with the pH increase in the wastewater from refinery R1, which means an increase of the emulsion stability and this can be attributed to the adsorption of hydroxyl ions at the oil-water interface. This indicates that the oil droplets are stabilized mainly by ionic surfactants present in the wastewater. The inverse tendency was observed in the wastewater from refinery R2, the removals increased gradually with the pH increase. Consequently the emulsion stabilization can be attributed basically to non-ionic substances in this case. The results showed also that the pH variation had very low effect of on the removals in the range pH of 6-8. That is why the test with the different coagulants and flocculants were performed at the natural pH of the wastewater. As it can be observed on Fig.5 a drastic increase of the COD removal was obtained at pH of 12. This can be attributed to the intense precipitation of Ca and Mg compounds which contribute to the emulsion destabilization. This phenomenon had a very strong effect in the effluent API-D1 which had the highest hardness and salinity.

The emulsion destabilization was obtained satisfactorily using combinations of mineral coagulant and polymers, as well as applying only cationic polymer of high molecular weight. The obtained results when using different mineral coagulants for the emulsion destabilization in the effluent API-D1 are illustrated on Fig.6. It can be observed that the polyaluminium chlorides had better behavior compared with the conventional coagulants. COD removals higher than 65% were reached with doses 30% lower than the required for the conventional coagulants. The best results were obtained with PAX-16S. Both aluminium and ferric sulphates proved to be effective destabilizing agents. The pH optimization tests indicated that the optimum pH for Al and Fe coagulants was 7.8 and 7.1 respectively. This is expected because the maximum neutralization of the oil droplets surface charge by hydrolyzed aluminium and ferric cations occurs in the pH range of 7-8 (Al-Shamrani et al., 2002). Similar optimal doses for each chemical product were obtained in the three studied effluents.

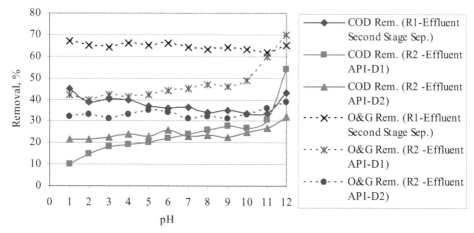

Fig. 5. Removals of O&G and COD before flocculation as a function of pH.

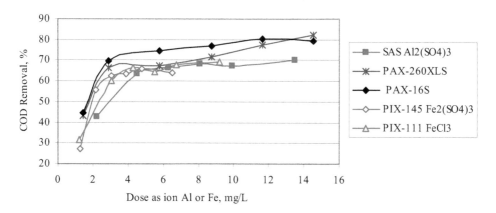

Fig. 6. Removals of COD using mineral coagulants (oily water with O&G, COD and TSS of 63-96, 503-566 and 65-74 mg/L respectively)

The removals obtained with the application of the different coagulants are summarized in Table 4. The results show that the addition of highly charged cations in the form of aluminium and ferric salts effectively induced the destabilization of the oil-water emulsions, leading to the significant oil separation (O&G and COD removal efficiencies of 61-79% and 61-70% respectively). TSS, turbidity and color were also successfully removed obtaining 69-85%, 92-97% and 87-89% efficiencies respectively. These results were expected, as the oil droplets have negative values of zeta potential (Nalco, 1995; Al-Shamrani et al., 2002). However, the flocs formed in the coagulation process were small in size and their settling was very slow. Therefore combinations of mineral coagulants with different polymers were tested. In these tests the coagulants were added at doses equal to 70% of the optimal doses indicated in Table 4. The results obtained in the effluent API-D1 are presented on Fig.7 and Fig.8. Both kinds of polymers, cationic and anionic ones, improved the COD removal. Lower COD concentrations were reached with the cationic polymers compared with the obtained with the anionic ones. The COD removals were calculated in the ranges of 78-93% and 66-81% for the cationic and for the anionic polymers respectively. The O&G removals were of 94-97% and 89-92% for the cationic and for anionic polymers respectively. The TSS removal was also better, efficiencies of 89-92% and 86-89% were obtained for the cationic and for anionic polymers respectively. Since the oil droplets are negatively charged, the better performance of the cationic polymers can be attributed to the increase of the cationic charge added to the oily wastewater, which enhances the reduction of the zeta potential and improves this way the destabilization of the oil-water emulsion. The anionic polymers combined with the mineral coagulants had only flocculating effect. The flocks formed in these tests were much greater and heavier than the obtained when only coagulants were used. The sludge quantities were of 40-60 ml/L.

The best coagulant-flocculant combinations and their optimal doses are summarized in Table 5. The O&G and COD removal efficiencies of 93-96% and 89-95% respectively were reached, which is almost 24% higher than the obtained using only coagulants. TSS, turbidity and color removal efficiencies were 81-90%, 99% and 94-97% respectively, that is 5-8% higher than the efficiency using only coagulant. The obtained in the performed tests removal efficiencies are higher than the reported by Galil & Wolf, 2001 and the determined optimal doses are lower than the reported in Galil & Rebhun, 1992.

Coagulant	Opti mal doses, mg/L	Removal efficiencies, %								
		R1-Effluent Second Stage Separator			R2-Effluent API-D1			R2-Effluent API-D2		
		O&G	COD	TSS	O&G	COD	TSS	O&G	COD	TSS
Aluminium sulphate (SAS)	50	62	67	83	62	63	69	61	62	76
PAX-XL60S	45	64	67	84	-	-	-	-	-	-
PAX-260XLS	30	-	-	-	64	66	80	66	67	78
PAX-16S	30	65	68	85	66	70	86	67	68	77
PAX-XL19	40	63	65	80						
Ferric chloride (PIX-111)	15	-	-	-	75	66	85	78	65	77
Ferric sulphate (PIX-145)	20	-	-	-	77	62	85	79	64	79
Ferric sulphate (Ferrix-3)	20	65	68	82	-	-	-	-	-	-

Table 4. Removals of O&G, COD and TSS obtained using only coagulants in the different API effluents (the doses are expressed in mg/L of chemical product)

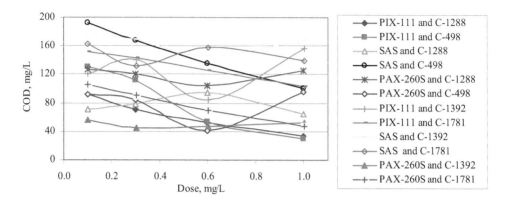

Fig. 7. Removals of COD using mineral coagulants and cationic polymers (oily water with O&G, COD and TSS of 96-120, 592-733 and 60-78 mg/L respectively)

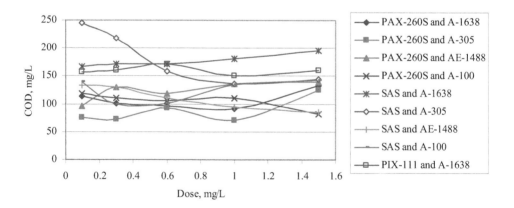

Fig. 8. Removals of COD using mineral coagulants and anionic polymers (oily water with O&G, COD and TSS of 96-110, 404-490 and 62-75 mg/L respectively)

Coa-gulant	Optimal doses, mg/L	Flocculant	Optimal doses, mg/L	Removal efficiencies, %								
				R1-Effluent Second Stage Separator			R2-Effluent API-D1			R2-Effluent API-D2		
				O&G	COD	TSS	O&G	COD	TSS	O&G	COD	TSS
SAS	45	ECOFLOC	0.4	96	93	87	-	-	-	-	-	-
PAX-260XLS	40	C-1288	0.6	95	91	88	-	-	-	-	-	-
PAX-260XLS	31	C-1392	0.3	-	-	-	96	94	85	93	92	84
SAS	35	C-1288	0.3	-	-	-	95	90	83	94	89	83
PIX-111	11	C-1288	1.0	-	-	-	93	95	81	93	93	83
PIX-145	14	C-498	1.1	-	-	-	96	95	90	94	93	88

Table 5. Removals of O&G, COD and TSS obtained using coagulants and flocculants in the different API effluents (the doses are expressed in mg/L of chemical product)

The results of the tests adding only cationic polymers for the emulsion destabilization and flocculation are presented on Fig.9. All studied polymers provided good COD, O&G and TSS removals, very similar to the obtained with coagulant and flocculant addition. The obtained COD, O&G and TSS removal efficiencies were of 81-94%, 83-96% and 78-95% respectively. The sludge generation adding cationic polymers was 20-30 ml/L, almost 50% lower than the obtained in the tests with the combinations of coagulant and polymers. The tests with pH variation indicated that the optimum pH was different for each polymer, the optimal pH values were in the range 6.9-8.5. The optimum pH were different for the three studied effluents. The removals obtained with the application of the different coagulants and the optimum pH values are summarized in Table 6. The flocculants ECOFLOC and C-1288 had the best performance for the oily effluent from the second stage separators of refinery R1 and C-5100 and C-1288 for both effluents of the refinery R2.

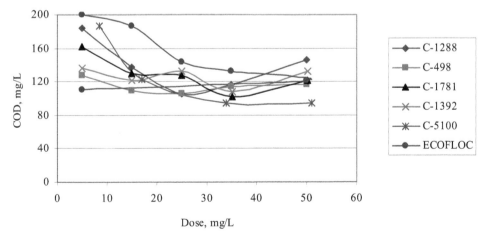

Fig. 9. Removals of COD using cationic polymers (oily water with O&G, COD and TSS of 142-164, 500-651 and 84-95 mg/L respectively)

Cationic polymers	Opti mal doses, mg/L	Opti mal pH	Removal efficiencies, %								
			R1-Effluent Second Stage Separator			R2-Effluent API-D1			R2-Effluent API-D2		
			O&G	COD	TSS	O&G	COD	TSS	O&G	COD	TSS
C-1288	30	7.4	94	85	94	-	-	-	-	-	-
C-1288	25	7.0	-	-	-	96	84	82	93	83	83
C-498	40	7.4	91	83	91	-	-	-	-	-	-
C-498	25	7.2	-	-	-	93	84	88	92	82	87
C-1781	35	7.2	92	85	93	-	-	-	-	-	-
C-1781	35	7.2	-	-	-	92	83	95	90	80	91
C-1392	40	7.2	92	83	90	-	-	-	-	-	-
C-1392	35	7.0	-	-	-	91	83	88	89	81	90
C-5100	34	7.6	-	-	-	95	94	91	92	90	92
ECOFLOC	30	7.4	95	86	95	-	-	-	-	-	-
ECOFLOC	50	7.2	-	-	-	83	81	78	82	82	80

Table 6. Removals of O&G, COD and TSS obtained using only coagulants in the different API effluents (the doses are expressed in mg/L of chemical product

The combination of processes flocculation and dissolved air flotation was first performed in refinery R1. The used oily wastewater had O&G, COD and TSS concentrations of 286 mg/L, 1,390 mg/L and 207 mg/L respectively. The saturation pressure (P) was varied from 40 to 70 lb/in², the recycling ratio (R) from 0.1 to 0.4. Both cationic polymers ECOFLOC and C-1288 were used in the tests. The results of the first experimental run indicated that the most important factor for the O&G, COD and TSS removal is the selection of the polymer, followed by the recycling ratio and finally the saturation pressure. ECOFLOC showed better performance than C-1288 in these tests. The effect of P and R variation on O&G concentration in the treated water using ECOFLOC is illustrated on Fig.10 (a). It can be observed that R values higher than 0.2 caused an increase of O&G concentration in the effluent. The increase of the O&G concentration was higher when high P values are applied. The values of the COD were between 111 and 309 mg/L. The determined O&C, COD and TSS removal efficiencies were of 74-99%, 78-92% and 73-89% considering all of the obtained results in this experimental run.

As the P reduction provided lower O&G concentrations in the effluent, the second experimental run considered P variation in lower range 35-55 lb/in² and R variation between 0.05 and 0.20. The initial O&G, COD and TSS concentrations varied in the ranges of 175-480 mg/L, 1,050-1,500 mg/L and 268-292 mg/L respectively. The effect of P and R variation on the O&G concentration in the treated water is illustrated on Fig.10 (b). The treated water O&G, COD and TSS concentrations were of 2-113 mg/L, 121-950 mg/L and 21-89 mg/L respectively considering all of the obtained results in this experimental run. ANOVA indicated that the most important factor for the O&G, COD and TSS removal was the recycling ratio, followed by a combined effect of R and the initial concentration. With minimum air/solid ratio of 0.10 a surface charge of 0.94-2.30 m³.m⁻².h⁻¹ was obtained in the flotation cell. According to the obtained optimization model O&G, COD and TSS removals more than 97%, 89% and 91% respectively can be obtained using low pressures in the saturation tank, of 37-40 lb/in², with 0.07-0.09 recycling ratio.

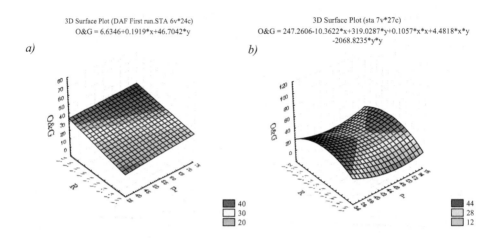

a) 3D Surface Plot (DAF First run.STA 6v*24c)
O&G = 6.6346+0.1919*x+46.7042*y

b) 3D Surface Plot (sta 7v*27c)
O&G = 247.2606-10.3622*x+319.0287*y+0.1057*x*x+4.4818*x*y
-2068.8235*y*y

Fig. 10. Effect of P and R variation on O&G concentration in the treated water: a) first experimental run; b) Second experimental run (refinery R1)

The third run flocculation-flotation tests were performed using the cationic polymer C-5100 and wastewater from the API effluent D1. The initial O&G, COD and TSS were 54, 414 and 120 mg/L respectively, much lower than the values in the previous tests. Color and Turbidity were 2,630 PtCo and 379 NTU. The P, R and HRT were varied in the ranges of 14-28 lb/in², 0.1-0.30 and 15-25 min. The effect of the HRT and R on the O&G concentration in the treated water using C-5100 is shown on Fig.11. The R had more significant effect on the removal of all the parameters compared with the one of the HRT. The best operational conditions were: P of 21 lb/in², HRT of 25 min and recycling ratio of 0.2. The obtained removal efficiencies for O&G, COD and TSS were 50-85%, 47-61% and 56% respectively. The Turbidity and Color removals were determined of 83-85% and 85-92% respectively.

3D Surface Plot (Spreadsheet3 .sta 15v*26c)
O&G = 14.8622-470.2667*x+4.401*y+987.6667*x*x
+3.78*x*y-0.1299*y*y

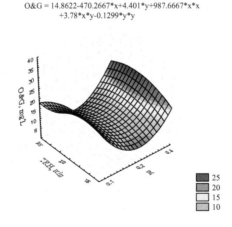

Fig. 11. Effect of HRT and R variation on O&G concentration in the treated water (R2)

The flocculation-floatation tests indicated that concentrations of O&G and TSS lower than 50 mg/L can be obtained in the treated oily wastewater. The O&G and TSS removal efficiencies were in agreement with the reported in Eckenfelder (2000) and Galil & Wolf, (2001), while the obtained COD removals were better than the reported in the literatuure. However, in spite of the obtained good COD removal efficiencies, the remaining values of the COD in the treated water were still high, in the range of 160-800 mg/L depending of the COD in the API effluents. These COD quantity, attributed basically to soluble organic matter, needs to be removed before the application of advanced treatment processes.

3.4 Evaluation of the biological and advanced treatment and analysis of the reclaimed water reuse feasibility

Different biological treatment processes have been used for refinery wastewater treatment, and the biological treatment systems allow good organic matter degradation; however, inhibition problems may occur because of the presence of many recalcitrant and toxic hydrocarbons, as for example the phenols. Biological treatment systems were already implemented in the studied refineries. For protection of the process performance, they have established maximum permissible limits (MPL) for some parameters which have to be accomplished in the influents to the biological reactors. The phosphate concentration in the refinery wastewater is generally low, so phosphoric acid is frequently used to support the biomass growth. As it could be seen in the previous Tables 1 and 2, ammonia nitrogen concentrations are relatively high in the wastewater and their removal is frequently an additional object of the biological treatment. Te refinery R1 has implemented sequential batch reactors (SBR) and refinery R2 a nitrification-denitrification activated sludge (AS) system. The refinery R1 has established the following MPL: 70 mg/L for O&G, 470 mg/L for COD, 30 mg/L for TSS, 6 mg/L for phenols, 560 mg/L for chlorides, 30 mg/L for sulfides and 6-9 units for pH. The MPL in refinery R2 are: 48 mg/L for O&G, 400 mg/L for COD, 5 mg/L for phenols, 560 mg/L for TDS, 8.5 units for pH and 35°C for temperature. The existing oily wastewater pretreatment facilities in the studied refineries normally accomplish these requirements due to their high retention capacity. However, the frequent operational problems made impossible the introduction of all the wastewater to the biological treatment systems. The reengineering project, considers design, construction and installation of new API separators and flocculation-DAF systems in both refineries. The obtained results of the treatability tests indicated that the suggested pretreatment systems provide the accomplishment of the established MPL for biological treatment. The averages of the physical-chemical parameters, obtained using one year operational data of the current biological systems are presented in Table 7. It has to be mentioned that the nitrification-denitrification AS reactor receive almost 30 L/s domestic wastewater which is treated in conjunction with the refinery effluent, while the SBR receive only pretreated refinery effluent. COD and NH_4-N removal efficiencies of 65% and 96% respectively were obtained in both biological treatment systems. As it can be expected nitrification-denitrification AS provided higher TKN removal compared with the SBR, 86% and 68% respectively. The O&G and phenol removals were also higher in the AS system. The average O&G removal efficiencies were 94% and 86% in AS and SBR respectively, and the phenol removals were 82% and 70% respectively. Sulphide removal efficiencies were of 95-96%.

Parameter	SBR in refinery R1		Nitrification-denitrification AS in refinery R2		Quality for cooling water make up
	Influent	Effluent	Influent	Effluent	
O&G, mg/L	50.1±8.4	7.2±3.4	48.6±10.3	3.2±1.8	-
COD, mg/L	453±74	157±42	388±82.5	137±35	75
TSS, mg/L	39.3±9.2	54.1±15.3	44.4±12.1	57.6±19	100
NH$_4$-N, mg/L	12.5±5.4	0.5±0.3	28.7±10.5	1.0±0.9	1
TKN, mg/L	20.3±3.4	6.4±2.5	45.2±5.3	6.2±2.4	-
Ptotal, mg/L	1.0±0.1	0.9±0.1	0.7±0.1	0.5±0.1	1
Phenols, mg/L	0.2±0.05	0.06±0.02	1.1±0.8	0.2±0.1	-
Sulphides, mg/L	11.3±7.1	0.5±0.3	27.0±10.3	1.3±0.8	-
Hardness, mg CaCO$_3$/L	397±52	385±44	253±44	238±24	650
Alkalinity, mg CaCO$_3$/L	125±12	105±33	103±27	84±15	350

Table 7. Performance of the biological treatment systems in the studied refineries

One of the reuse options in the refineries is in the cooling tower make-up. The most frequent water quality problems in cooling water systems are scaling, corrosion, biological growth, foaming, as well as fouling in the heat exchangers and condensers. To avoid these potential problems the reclaimed water used in cooling systems must not supply nutrients (nitrogen and phosphorus) or organics that promote the growth of biofilms. The cooling water must no lead to the formation of scale (calcium and magnesium precipitation). Mexican refineries use recirculating cooling systems and the quality of the water input is of big concern. The comparison of the water characteristics of the secondary effluents (Table 8) with the quality requirements for cooling water make-up (US EPA, 1980) indicated that the concentrations of TSS, NH$_4$-N, P, Hardness and Alkalinity were lower than the suggested ones. The effluent COD was higher than the suggested value, however the organic matter present in the secondary effluent is constituted basically of compounds difficult for biodegradation. This organic matter could difficultly improve the biofilm growth. Therefore the secondary effluent could be used in the cooling system. Generally, all water reuse for cooling water make-up uses a lime clarification process prior to reuse (US EPA, 1989). This process can reduce hardness, phosphates, silicates and colloidal organics. Filtration is frequently recommended after the liming process.

Advanced treatment facilities were already implemented in the studied refineries. The refinery R1 has lime softening (LS) reactors; pressure sand filters (F) and reverse osmosis (RO) system. Lime, soda ash and flocculant are used in the liming process. The effluent is acidified and then passed through the filters. Then the filtered effluent is submitted to a chlorine disinfection and send to a storage tank for its reuse in the cooling system of the refinery. Part of this water (86 L/s) is directed to the reverse osmosis system and the desalted water is supplied for use in different oil refining processes. The characteristics of the effluents from the advanced treatment processes are presented in Table 8. As it can be seen the liming process followed by filtration enhanced the water quality of the secondary effluent. This treatment allowed Hardness removal of 80%, TSS removal of 74%, Si removal of 71%and a complete P, O&G, S^{2-} and phenol removals. Additionally, COD removal efficiency of 48% was obtained and this way COD was reduced to 86 mg/L after liming and filtration. The obtained water quality is perfectly proper for their reuse in the cooling system. The reverse osmosis system provided 93% TDS removal and 82% hardness removal. COD was reduced to 31 mg/L and the rest of the characteristic pollutants were not detected in the effluent. The obtained water quality allows the use of the RO effluent in most of the production processes.

	Advanced treatment in R1				Advanced treatment in R2		
	Influent	LS	F	RO	Influent	UF	RO
O&G, mg/L	7.2±3.4	2.1±0.2	ND	ND	3.2±1.8	ND	ND
COD, mg/L	157±42	97±33	86±31	31±7	137±35	92±32	28±5
TSS, mg/L	54±15	45±8	14±6	ND	57±19	2±1	ND
NH4-N, mg/L	0.5±0.3	0.4±0.1	0.4±0.1	ND	1.0±0.9	1.0±0.2	ND
TKN, mg/L	6.4±2.5	5.3±1.8	5.1±1.1	ND	6.2±2.4	1.0±0.3	ND
Ptotal, mg/L	0.9±0.1	ND	ND	ND	0.5±0.1	0.3±0.1	ND
Phenols, mg/L	0.06±0.02	ND	ND	ND	0.2±0.1	0.1±0.1	ND
Sulphides, mg/L	0.5±0.3	ND	ND	ND	1.3±0.8	0.2±0.1	ND
TDS, mg/L	779±56	683±61	668±49	46±15	876±38	780±40	54±11
Hardness, mg CaCO3/L	385±44	75±23	67±22	12±3	238±24	221±19	39±7
Alkalinity, mg CaCO3/L	105±33	86±27	83±22	23±4	84±15	83±14	23±5
Si, mg/L	21±3	6±5	6±5	ND	24±4	23±4	14±2

ND-Not detected

Table 8. Performance of the advanced treatment systems in the studied refineries

The refinery R2 has ultrafiltration and inverse osmosis systems. The effluent from the nitrification-denitrification AS system is disinfected and then submitted to ultrafiltration. Most of the obtained effluent is directed to the cooling system of the refinery. Part of the filtered water (50 L/s) is submitted to reverse osmosis desalting and after that reused in the refining processes. The water characteristics after each treatment process are presented in Table 8. The ultrafiltration removed basically the TSS and O&G, as well as reduced the COD values and the concentrations of the rest of the water quality parameters. The effluent of the UF can be used in the cooling systems. The reverse osmosis allowed the same removals of TDS, hardness and COD as the one in the refinery R1 and the obtained water can be used in the production processes.

The performed analysis demonstrates that both refineries have the capacity to obtain two water qualities reclaimed water for reuse. The reengineering of the existing pretreatment systems will ensure the obtaining of water with proper quality to be submitted to the existing already biological and advanced treatment. This way all the wastewater could be treated and that will make feasible the implementation of the proposed new water management options in both refineries.

4. Conclusions

Petroleum refining industry has a high potential for implementation of water conservation strategies. After a suitable treatment, the totality of the petroleum refining wastewaters can be reused, obtaining therefore the protection of the receiving water bodies and reducing the fresh water demand. The performed study of the water management systems in two refineries allowed the development of alternatives which could provide fresh water savings of 51-59%. It is possible to obtain high quality treated water not only for reuse in the cooling towers but also for the production processes and auxiliary services. The pretreatment of the oily wastewaters using primary oil gravitational separators and chemically enhanced separation processes allows a successful implementation of biological treatment, followed by advanced processes. The use of reclaimed municipal wastewater in the cooling towers make-up allows further fresh water saving opportunity. The waste management has to consider separate treatment of sour waters and for the spent caustics, as well as a pretreatment of all effluents whose main pollutants are oil, solids and sulfides. Cleaner production actions have to be implemented for the reduction of the pollutants in the wastewater.

The preliminary separation of the free oil by natural flotation allows 90-95%O&G removal efficiency with surface loading rates of 1.15-4.60 $m^3.m^{-2}.h^{-1}$. As the floatation velocity of the oil droplets depend of the oil characteristics which are different for each refinery, the performance of experimental tests are highly recommended for the obtaining of reliable design parameters. The TSS and COD removals obtained in the performed treatability tests were of 62-72% and 34-39%. The increase of the hydraulic retention time in the range 0.5-2.0 h improves the TSS and COD removal in the separators. The effluents from the separators had low O&G concentration (47-62 mg/L), however the remained COD was higher than 340 mg/L. The further O&G and COD removal requires emulsion destabilization followed by separation process. The emulsion destabilization can be reached using combinations of mineral coagulants and polymers, as well as applying only cationic polymers of high molecular weight and high charge density. The addition of highly charged cations in the form of aluminium and ferric salts effectively induced the destabilization of the oil-water emulsions. Similar behavior was obtaining with Fe and Al salts. Polyaluminium chlorides had better behavior compared with the conventional coagulants. COD removals higher than 65% were reached with doses 30% lower than the required for the conventional coagulants. The combinations of mineral coagulants with cationic polymers provided O&G and COD removal efficiencies of 93-96% and 89-95% respectively, which is almost 24% higher than the obtained using only coagulants. Similar results were obtained applying only cationic polymers and the generated sludge was almost 50%lower than the generated with the combinations of coagulant y polymers. The characteristics of the oil-water emulsion may be different in each refinery. Therefore, the selection of the best chemical product for the emulsion destabilization, as well the determination of the optimal doses and pH, are crucial for the process success. The combination of flocculation and dissolved air flotation provides good O&G, COD and TSS removal efficiencies. Concentrations O&G and TSS lower than 50 mg/L can be obtained in the effluent. The COD removals vary in the range 47-92%. The experimental tests demonstrated that the most important factor for the O&G, COD and TSS removal is the selection of the polymer, followed by the recycling ratio. The effect of the saturation pressure, the hydraulic retention time were lower. The best results were obtained with relatively low pressures of 21-40 lb/in^2 and recycling ratio of 0.1-0.2. In spite of the obtained high COD removals, the remaining values in the treated water are still high. These

COD quantity, attributed basically to soluble organic matter, has to be removed before the application of advanced treatment processes.

The performed evaluation of two real scale biological treatment systems, sequential batch reactors (SBR) and nitrification-denitrification activated sludge (AS) system showed COD and NH_4-N removal efficiencies of 65% and 96% respectively were obtained in both cases. Nitrification-denitrification AS provided higher TKN removal compared with the SBR, 86% and 68% respectively. The O&G and phenol removals were also higher in the AS system. The average O&G removal efficiencies were 94%and 86% in AS and SBR respectively, and the phenol removals were 82% and 70%respectively. Sulphide removal efficiencies were of 95-96%. The secondary effluents accomplish the required water quality for reuse in cooling system make-up. For better TSS control and additional enhancement of the secondary effluent water quality, filtration or ultrafiltration can be recommended. Lime softening of the secondary effluent can be implemented before filtration if the hardness of the wastewater is higher than the established limit for reuse or when the reverse osmosis system design establishes restrictions with respect of the Hardness in the water to be demineralized. The last one was the case of refinery R1. The obtaining of the second water quality of water for reuse in production processes is technically feasible using reverse osmosis systems.

5. References

Al-Shamrani, A.A., James, A. & Xiao, H. (2002). Destabilisation of oil–water emulsions and separation by dissolved air flotation. *Water Research*, Vol. 36, No.6, pp.1503–1512.

API, American Petroleum Institute. (1990). *Design and operation of oil-water separators. API Publication*. Washington D.C.

Baron, C., Equihua, L.O. & Mestre, J.P. (2000). B.O.O.Case: water manajement project for the use of reclaimed wastewater and desalted seawater for the "Antonio Dovali Jayme" refinery, Salina Cruz, Oaxaca, Mexico. *Water Science and Technology*, Vol. 42, No.5-6, pp.29-36.

Daxin Wang, Flora Tong & Aerts P. (2011). Application of the combined ultrafiltration and reverse osmosis for refinery wastewater reuse in Sinopec Yanshan Plant. *Desalination and Water Treatment, Vol.25, No.1-3, pp.133–142.*

EC (European Commission). (2000). *Integrate pollution prevention and control: Reference document on best avaible technologies in common wastewater and waste gas*, Institute for Perspective Technological Studies, Seville.

Eckenfelder, W.W. (2000). *Industrial Water Pollution Control*, 3rd.ed., McGraw-Hill.

Elmaleh S. & Ghaffor N. (1996) Upgrading oil refinery effluents by cross-flow ultrafiltration. *Water Science and Technology,* Vol.34, No.9. pp. 231–238.

Farooq, S. & Misbahuddin, M. (1991). Activated carbon adsorption and ozone treatment of a petrochemical wastewater. *Environmental Technology*, Vol.12, No.2, pp.147-159.

Levine, A.D. & Asano T. (2002). Water reclamation, recycling and reuse in industry. In: *Water recycling and resource recovery in Industry*, Editted by P. Liens, L. Hulshoff Pol, P. Wilderer and T. Asano, IWA Publishing, p.29-52.

Galil, N. & Rebhum, M. (1992). Waste management solutions at an integrated oil refinery based on recycling of water, oil and sludge. *Water Science and Technology*, Vol.25, No.3, pp.101-106.

Galil, N. & Wolf, D. (2001). Removal of hydrocarbons from petrochemical wastewater by dissolved air flotation. *Water Science and Technology*, Vol.43, No.8, pp.107-113.

Guarino C. F., Da-Rin B. P., Gazen A. and Goettems E. P. (1988). Activated carbon as an advanced treatment for petrochemical wastewaters. *Water Science and Technology*, Vol.20, No.10, pp. 115-130.

IPIECA (International Petroleum Industry Environmental Conservation). (2010). *Petroleum refining water/wastewater use and management.* Operations Best Practice Series, London, UK.

Lee, L.Y, Hu, J.Y., Ong, S.L., Ng, W.J., Ren, J.H. & Wong, S.H. (2004) Two stage SBR for treatment of oil refinery wastewater. *Water Science and Technology*, Vol.50, No.10, pp.243-249.

Misković, D., Dalmacija, B., Živanov, Ž., Karlović, E., Hain, Z. & Marić S. (1986). An investigation of the treatment and recycling of oil refinery wastewater. *Water Science and Technology*, Vol.18, No.9, pp.105-114.

Mukhetjee, B., Turner, J. & Wrenn, B. (2011). Effect of oil composition on chemical dispersion of crude oil. *Environmental Engineering Science*, Vol. 28, No.7, 497-506.

Nalco Chemical Company (1995). *Manual del Agua. Su naturaleza, tratamiento y aplicaciones.(The Nalko Water Handbook)*, Tomo I, II, III. Segunda edición. McGraw-Hill/Interamericana de México, S.A. de C.V.

PEMEX (Mexican state-owned petroleum company). (2007). *Principales estadísticas operativas (Basic operation statistics)*, México D.F..

Powel, S. T. (1988). *Manual de aguas para usos industriales.* Vol. 1, 2, 3. Primera reimpresión, Ediciones Ciencia y Técnica, S.A. de C.V., México, D.F.

Schneider, E.E., Cerqueira, A.C.F.P. & Dezotti, M. (2011). MBBR evaluation for oil refinery wastetreatment with post-ozonation and BAC, for water reuse. *Water Science and Technology*, Vol. 63, No.1, pp.143-148.

Standard Methods for the Examination of Water and Wastewater. (2005). 21th edition, American Public Health Association/American Water Works Association/Water Environment Federation, Washington DC, USA.

Sastry, C A. & Sundaramoorthy, S. (1996). Industrial use of fresh water vis-a-vis reclaimed municipal wastewater in Madras, India. *Desalinisation*, Vol.106, pp.443-448.

Teodosiu, C.C., Kennedy, M. D., van Straten, H.A. & Schippers, J.C. (1999). Evaluation of secondary refinery effluent treatment using ultrafiltration membranas. *Water Research*, Vol.33. No.9, pp.2172-2180.

US EPA (U.S. Environmental Protection Agency). (1982). *Development Document for Effluent Limitations Guidelines and Standards for the Petroleum Refining Point Source Category*, Washington, D.C.

US EPA (U.S. Environmental Protection Agency). (1980). *Treatability manual, EPA 600/8-80-042E*, Vol. 1, 2, 3, 4, 5. Washington, D.C.

US EPA (U.S. Environmental Protection Agency) and US AID (US Agency for International Development). (1980). *Guidelines for Water Reuse, EPA 625/R-92/004*, USA.

US EPA (U.S. Environmental Protection Agency). (1995). *Profile of the Petroleum Industry.* EPA/310-R-95-013. Washington, D.C.

WB (World Bank). (1998). *Pollution Prevention and Abatement Handbook: Petroleum Refining*, Technical Background Document, Environment Department, Washington, D.C.

WEF (Water Environment Federation). (1994). *Pretreatment of industrial wastes.* Manual of Practice FD-3, Alexandria, USA.

Zubarev, S. V., Alekseeva, N. A., Ivashentsev, V. N.,Yavshits, G. P., Matyushkin, V. I., Bon, A. I. & Shishova, I. I. (1990). Purification of wastewater in petroleum refining industries by membrane methods. *Chemistry and Technology of Fuels and Oils*, Vol.25, No.11, pp.588-592.

8

Determination of the Storage Volume in Rainwater Harvesting Building Systems: Incorporation of Economic Variable

Marina Sangoi de Oliveira Ilha[1] and Marcus André Siqueira Campos[2]
[1]Department of Architecture and Construction, School of Civil Engineering,
Architecture and Urban Design, University of Campinas, Campinas, SP,
[2]School of Civil Engineering, Federal University of Goiás, Goiânia, GO,
Brazil

1. Introduction

Rainwater harvesting has been used as a technique to promote water conservation in buildings, as it substitutes the potable water in activities where the use of potable water is not required.

In spite of the surge in interest over recent years, some questions still remain regarding to these systems, mainly what involves the reservoir sizing. There are many methods for this purpose that use different inputs such as: rainwater demand, catchment area, roof material, rainy data (daily or monthly) and dry periods. Even in the Brazilian Standard (ABNT, 2007), there is no consensus as to which method should be used. Table 1 shows the main methods found in the literature and their respective inputs.

Mainly in developing countries, actions that promote water conservation must be economically feasible so it can raise the interest in investments. Moreover, urban lots are progressively smaller and more expensive. These variables can restrict the size of the reservoirs used in a rainwater system and this should be considered in their design.

This article proposes the use of an optimization technique to find the most adequate volume of rainwater reservoirs i.e. the optimal economical result measured by the Net Present Value (NPV): the Particle Swarm Optimization (PSO).

PSO is a population-based technique of stochastic nonlinear functions. Its use was inspired by social behavior in flocking birds or school of fishes (Boeringer, Weiner, 2004). It was used for this optimization process because of its flexibility and because it allows the inclusion of other variables that might interfere with the NPV calculation in any given future. This aspect expands the capacity of data processing without loss of efficiency of the algorithm.

In this study, PSO was used to size rainwater reservoirs in four case studies and the results obtained were compared with traditional methods that have been used for this purpose, verifying the improvement of the decision making process.

SIZING METHOD	Source	Annual rainfall	Monthly rainfall	Daily rainfall	Catchemnt area	Annual Demand	Montly Demand	Daily Demand	Roof Material
Annual Average	Gould; Nissen- Pettersen (1999)	x			x				
Brazilian Pratical Method	ABNT (2007)	x			x				
English Practical Method	ABNT (2007)	x			x				
German Practical Method	ABNT (2007)	x				x			
Australian Practical Method	ABNT (2007)		x			x			x
Rippl (Monthly data)	Thomas (2003); Campos (2004); ABNT (2007); Yruska (2010);		x				x		x
Rippl (Daily data)	Thomas (2003); Campos (2004); ABNT (2007); Yruska (2010);			x				x	x
Netuno@	Guisi et al(2007); Rocha (2009)					x			
Numerical Simulations	Fewkes (1999); Liao et al (2005); Liaw; Tsai (2004)				x	x		x	x
Weibull	Group Raindrops (2002); Simioni et al (2004)					x		x	

Table 1. Reservoir Sizing Methods and Inputs

2. Particle swarm optimization

The PSO algorithm is very similar to other evolutionary algorithms such as genetic algorithms (GA): the system takes a starting point with a population of variables and then research is done to find optimal solutions by the updating of generations. However, unlike the GA, there are no evolution operators, such as crossovers or mutations. Potential solutions, here called "particles", fly over the space of the problem, following the best particles (Particle Swarm Optimization, 2009).

An individual (particle) in communities as flocks or schools learns not only with the experiences that it had, but also with the experiences of the group to which it belongs. Thus, this technique tends to provide the best personal experience (position visited) and the best group experience.

The particles of PSO have a similar behavior. Through a simulation in a two-dimensional space, the velocity vector defines the displacement of the particle and another vector defines the position. The equations of these vectors are (Carrilho, 2007):

$$p_{k+1}^i = p_k^i + v_{k+1}^i \qquad (1)$$

$$v_{k+1}^i = \omega v_k^i + C_1 rand_1 \left(b_k^i - p_k^i \right) + C_2 rand_2 \left(b_k^g - p_k^i \right) \tag{2}$$

Where:

k - an increase in pseudo-time unit;

k_i - position of each particle i (candidate solutions) in time k (iteration);

k_{i+1} - position of the particle i at time k +1;

b_{ki} - best position reached by the particle i at time k - best individual position;

b_{kg} - best position of the swarm at time k- is the best position reached by a particle used to guide the other particles in the swarm;

v_{ik} - speed of the particle i at time;

kv_{ik+1} - set speed of the particle i at time k +1;

rand1 and rand2 - independent random numbers (with uniform probability) between 0 and 1.

C1 and C2 - control information flow between the current swarm: If C2 > C1 – particle swarm will place confidence in the swarm, otherwise it puts confidence in itself. C1 and C2 are known as cognitive and social parameters respectively.

ω - inhere factor (or damping factor), which controls the impact of previous velocity of the particle on its current speed.

There are many different fields of application for PSO. Wang et al (2009) investigated the feasibility of the PSO algorithm to estimate the quality parameters of a water body. From the results obtained, it was observed that the proposed algorithm provides satisfactory results, either in relation to the genetic algorithm also developed for this purpose, or in the control data. The authors concluded that it is an important tool for calibrating water quality models. Another use of the PSO algorithm is for planning water supply systems (Yang; Zhai, 2009; Montalvo *et at*, 2010). Yang, Zhai (2009) compared the results obtained with the application of a genetic algorithm and PSO, demonstrating the flexibility of PSO, enabling the adaptability of the optimization of discrete and continuous variables.

3. Methods

The present study consists of theoretical research which involves the following steps: Survey of the methods that is regularly used in Brazil to size rainwater reservoirs, application of those methods in four case studies, simulation of sizing considering such methods, and the analysis of results; proposition of a tool to determine the volume based reservation.

The development of the PSO Tool involved:

a. Cost Estimation of each reservoir: The costs of the fiberglass tanks were obtained in building material stores; and a local construction company gave the estimated costs for the concrete tanks. From this, functions were created for the estimation of the costs of the tanks:

$$C = 0.1733V + 32.927 \text{ (Fiberglass tanks)} \tag{3}$$

$$C = 0.4672V + 12.791 \text{ (Concrete tanks)} \tag{4}$$

Where:

C – Cost of the tank (R$; US$1.00=R$1.66)

V – Volume of the tank (liters)

b. Modeling of the water price policy – functions for the estimation of the tariff were used, based on the values and classes of consumption by SANASA (Local water company). For commercial buildings, these functions are:

$V \leq 10$ m³	$P = 32{,}50$	(5)
$10 < V \leq 20$ m³	$P = 5{,}42V - 21{,}70$	(6)
$20 < V \leq 30$ m³	$P = 8{,}63V - 85{,}90$	(7)
$30 < V \leq 40$ m³	$P = 10{,}15V - 131{,}50$	(8)
$40 < V \leq 50$ m³	$P = 11{,}82V - 198{,}30$	(9)
$V > 50$ m³	$P = 14{,}25V - 319{,}80$	(10)

Where:

V – water consumption (m³)

P - water tariff (R$; US$1.00 = R$1.66). The water tariff increase in the last 10 years was considered to calculate the average, maximum and minimum values for the simulations.

c. Determination of the Net Present Value (NPV) function

d. Use of PSO technique for optimizing the NPV function for each volume estimated.

The PSO based approach suggested in the present work aims to establish the optimal storage volume in a given rainwater harvesting building system, with regards to the maximization of the system's NPV. The system has two distinct modules: *simulation* and *optimization*.

The simulation module calculates the system's NPV over time, given a series of precipitations and tariff rates based on previous data. The simulation module's output is final NPV to be utilized as objective function.

The optimization module is based on a PSO in its version with global topology (*gbest* or *global Best PSO*). As previously described, the PSO is a search/optimization technique based on swarm intelligence, where the position of each particle in the search space represents a possible solution to the problem. In the suggested approach, the position of the particle in a given instant represents a possible storage volume for the system with the minimum volume (v_{min}) determined by the user and maximum (v_{max}) defined by the building occupation rate and the storage's maximum height. For the purposes of the experiment described here, the occupation has been set as 0,05% and the maximum height as 3m.

Initially, a 10 particle swarm was created and distributed uniformly in the search space on the interval $[v_{min}, v_{max}]$. Then, the *fitness* of each particle was calculated and for each one its *pbest* updated to its initial position. After that, *gbest* was defined as the position of the particle with the best *fitness* in the swarm. In the following iterations, the particles update their velocities according to the equation:

$$v_i(t+1) = v_i(t) + c_1 r_1(t)[y_i(t) - x_i(t)] + c_2 r_2(t)[y(t) - x_i(t)] \tag{11}$$

where $v_i(t)$ is the velocity of the particle in the instant t; $x_i(t)$ is the position of the particle i in the instant t, c_1 e c_2 are the acceleration constants that represent the social and cognitive components of learning and $r_1(t)$ e $r_2(t)$ are random values sampled from a uniform distribution $U(0,1)$. These values have the objective of introducing a stochastic element in the algorithm. In the experiments, the learning factors c_1 e c_2 were defined as 2. This value was obtained empirically, establishing a satisfactory balance between search capability and depth and width.

The best position found by a particle i so far (i.e., *pbest*) is represented by y_i. As this is a problem of NPV optimization, *pbest* is calculated as follows:

$$y_i(t+1) = \begin{cases} y_i(t) & if \quad f\big(x_i(t+1)\big) \le f\big(y_i(t)\big) \\ x_i(t+1) & if \quad f\big(x_i(t+1)\big) > f\big(y_i(t)\big) \end{cases}$$

Where $f : R \to R$ is the *fitness* function, represented as the NPV as function of the system's storage volume. If in a given instant t a particle x finds a position that produces a better NPV than any previously found, its *pbest* is updated to the position of this particle in the instant t.

On the other hand, the development of the case studies involved the following activities:

a. Building selection: two aspects were considered in this selection - the building location should be close to the University of Campinas, where the rainfall data were captured and, and all design data should be readily available;

b. Rainwater demand estimation: rainwater was considered for supplying the following non-potable uses: toilet flushing; landscape irrigation and floor washing. Six scenarios of rainwater use were constructed: only for close-coupled toilet flushing (BD), only for landscape irrigation (R), only for floor washing (L) and four combinations of these scenarios: BD+R, BD+L, R+L and BD+R+L;

c. Rainfall volume estimation: the period for the analysis of rainfall data was from January 1971 through June 2009. Daily and monthly averages and maximum daily rainfall intensity, periods of drought and their frequencies were also analyzed;

d. Selection of the methods for the determination of the reservation volume: the following methods were chosen, based on the literature survey: Rippl (using daily and monthly rainfall data); Weibull, Netuno®, and the practical methods recommended in the Brazilian Standard: Azevedo Neto, English; Australian and German;

e. Sensitivity analysis based on different lifetimes and tariff value. There is no reference for lifetime of these components in the literature investigated. Thus, a period of 20 years was estimated for concrete tanks and 10 years for fiberglass tanks. For the water tariff, adjustments made by the local water company were considered with the starting point being the implementation of the Real (1994) by 2009;

f. Completion of the simulation, using the tool developed in this study.

An overview of the decision making process is shown from the results obtained, with a) the "conventional" sizing method and sensitivity analysis and b) with the results of the simulation. The sensitivity analysis provides a large number of options and outcomes to assess the volume and demand that will offer the greatest financial return, measured by the NPV of each situation.

The results were compared and analyzed in both the quantitative and qualitative aspects: optimal volume, initial investment, and payback of the investment, efficiency, lot occupation, and ease of use of the model including the input data. This analysis was made to verify the feasibility of using the PSO as a tool that can improve the decision making process in the design of the rainwater system, taking crucial factors for the decision process into account.

4. Results

4.1 Development of the PSO tool
Figure 1 shows the flowchart for the PSO tool. This flowchart was used to develop the RAIN TOOLBOX® software. As mentioned earlier, the PSO technique was chosen for this optimization process because of its flexibility, which allows the inclusion of other variables that may have an impact on the future NPV calculation, expanding the capacity of data processing, optimizing other variables besides the volume, such as the position of the reservoir, treatment required, etc., without losing efficiency of the algorithm. The PSO was shown to be a fast technique: the results were obtained in few seconds. The processing speed depends on both the number of particles (volume) and the number of interactions. This software allows choosing these variables.

Figure 2 shows the interface of the RAIN TOOLBOX® software. The first version is in Portuguese, the English version is being developed. In square 1, the following input data is required: Total area of the lot, catchment area, and rate of the lot will be used for the tank and the runoff coefficient. Square 2 contains the input data concerning to costs of implementation and maintenance (monthly, bimonthly, semi-annual and annual). The material of the reservoir, the consumer class (to define the water tariff), the daily demand of rainwater, and the maximum height of the reservoir are input in dialog box 3. Box 4 requires the rainfall data and the historical water tariff adjustment to be input. Lastly, in Box 5, the number of particles and interactions along with the minimum volume to be searched is typed.

4.2 Case studies
The rainfall data of the studied region is characterized by a dry season, with long periods of drought with an onset in April extending until August and a rainy season, from September to March. In the period studied (1971-2009), the rainiest month was January, with 272mm yearly average, followed by December (236mm/month) and February (193mm/month). Yearly, in the aforementioned period, the rainiest year was 1983 (2619mm), and driest was 1978 (811mm).

4.2.1 Case 1 – Residential building
This case features a two-story building with two bedrooms, one with a suite (room with a bathroom) and a restroom on the upper floor. Downstairs, it can be found a kitchen, a laundry room, the living room and a bathroom. The house was designed to accommodate 5 people.

The lot is 450 m², with the building covering 160 m². The building is covered with ceramic roof tiles and it has two roof surfaces. The yard is approximately 150 m². The predicted use of rainwater is for irrigation in the yard and toilet flushing. It was supposed that the yard is irrigated once a week, using 1 liter/ m², always from 06:00h to 08:00h.

It is estimated that each inhabitant flushes 6 times a day, 4 times being liquid and twice solid waste. Thus, we have a total of 30 instances of use, 20 with partial volume and 10 with total volume. Through previous observation, a daily distribution pattern was estimated.

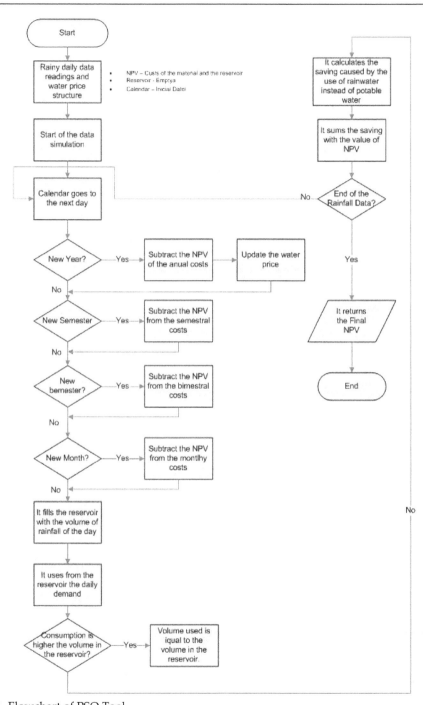

Fig. 1. Flowchart of PSO Tool.

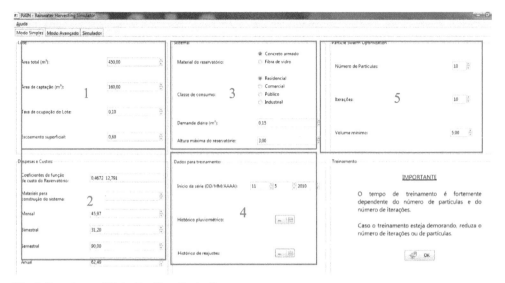

Fig. 2. Interface of Rain Toolbox® - in Portuguese.

The volume used by the toilets is 136 liters per day. Considering the 150 liters utilized in the yard's weekly irrigation, we have a total consumption of 1102 liters a week. Over 4 weeks (28 days), it was estimated that the demand for February is 4408 liters. For 31-day months a 1.107143 correction factor was applied and for 30-day months, a 1.071429 factor was applied, the result is, respectively, 4880.29 liters and 4722.86 liters.

Table 2 presents the reservation volumes obtained with the aforementioned methods.

Method	Reserved Volume (m³)	Efficiency (%) determined according to Campos (2004)	Efficiency (%) determined by Netuno Software
Rippl Monthly	1,00	53	63
Rippl Daily	1,85	65	76
Practical Brazilian	33,55	100	100
Practical English	11,96	98	98
Practical German	3,45	76	85
Practical Australian	1,00	53	63
Weibull's Method	7,29	90	94
Netuno Software	3,50	76	85

Table 2. Reservation volumes obtained with standard methods e by Netuno Software – Case Study 1 – residential building

Analyzing the obtained results, a considerable discrepancy can be seen in the results from the Brazilian and English practical methods that yielded unexpectedly high values considering the magnitude of the building. The other methods yielded reasonable results, all feasibly applicable in a residence; nevertheless, with this information, it is still hard to determine which value to use. Thus, it was decided that a sensibility analysis of the results was to be made, with economic performance as criterion, which is also this work's main purpose. Each result presented in Table 3 was analyzed in terms of its economic efficiency of investment, according to the flowchart in Picture 2.

It's important to consider that the initial investment consists solely of the cost of storage, as all other costs are fixed, independently of the volume of the storage.

The costs were estimated for concrete and glass fiber storages. To estimate the cost of the storages, the previously explained model was utilized.

According to the estimated potable water demand (200 l/hab.day), the potable water economy would be 4.88 m³, or U$10.84 monthly. However, as efficiency varies from volume to volume, this value will be proportional to its volume. The operating and maintenance cost was divided as follows: energy consumption – 30 working minutes per day: US$13.43/month; chlorine for purification - 4 g/m³: US$0.03/month; cost of the analysis according to the Brazilian Standard: chlorine and pH – US$0.43/month (using test strips); turbidity – US$7.23/month; color – US$7.23/month; total coliforms: US$27.10 once a semester; fecal coliforms: US$27.10 once a semester; system maintenance: cleaning of the storage, gutters and pump – a domestic worker's daily wage – US$37.59/year; cleaning of the filter – half a domestic worker's daily wage – US$37.59/year.

The monthly cost, based on once a semester and twice a semester, proportionally accounted for US$49.86, which is higher than what would be saved in the best possible scenario for a household (with 100% efficiency, US$10.87 would be saved monthly). Thus it can be concluded that, economically, the investment would never return. However, there are other factors, economics aside, that should be taken into account, such as the real value of water and other environmental advantages.

So, even without economic advantages it is possible to choose a rainwater harvesting system due to its environmental advantages. The chosen system, however, must be the least economically disadvantageous. Table 3 presents the determined NPV values for each of the aforementioned methods, as function of maximum, minimum and average adjustments of the water tariff, which are respectively: 19.58%, 5.60% and 10.89%/year.

To apply the Rain Toolbox to case study 1, the height of the storage was limited to 3.00m and it was established that it must occupy 5% of the terrain's total area. The simulation, using 10 particles and 10 iterations, yielded 3.00 m³ as result. For the concrete storage, the NPV was US$289.45 and for the fiberglass storage, it was US$5795.66. It was observed that the volume determined by the software was the same as the minimum posited (in this case 3.00 m³ was utilized to supply the daily demand).

What had already been shown was confirmed by traditional analysis; the costs (construction, operation and maintenance) for the system in such residences are higher than the returns: independently of the utilized volume there will be loss, and the lower the volume, the lower the loss.

4.2.2 Case 2 – Institutional building

This case features an institutional building consisting of a group of classroom buildings of the Faculty of Civil Engineering, Architecture and Urbanism of the State University of Campinas.

Method	Volume (m³)	Minimum Adjustment (5.60%)		Average Adjustment (10.89%)		Maximum Adjustment (19.58%)	
		20 years (concrete)	10 years (fiber)	20 years (concrete)	10 years (fiber)	20 years (concrete)	10 years (fiber)
Rippl Monthly	1.00	-2970.00	-2470.28	-2787.27	-2395.44	-2197.27	-2226.84
Rippl Daily	1.85	-31075.60	-2473.69	-2888.83	-2382.23	-2182.55	-2176.20
Practical Brazilian	33.55	-11576.60	-5485.16	-11240.10	-5344.46	-10153.50	-5027.49
Practical English	11.96	-5608.31	-3278.75	-5278.52	-3140.87	-4213.69	-2830.23
Practical German	3.45	-3450.34	-2559.99	-3194.59	-2453.06	-2368.80	-2212.16
Practical Australian	1.00	-2970.00	-2470.28	-2787.27	-2395.44	-2197.27	-2226.84
Weibull	7.29	-4386.74	-2855.34	-4083.87	-2728.72	-3105.96	-2443.44
Netuno	3.50	-3464.21	-2565.13	-3208.45	-2458.20	-2382.67	-2217.30

Table 3. NPV values (in US$) for each method and adjustment – Case Study 1 – residential building

It is comprised of three blocks, two already finalized and one still under construction. It has a total area of 1500 m², with four pavements. The total area for rainwater harvesting is 1222 m², covered by metallic roof tiles. Each pavement has two restrooms, each with 5 sinks and 5 close-coupled toilets and the men's restrooms have metallic gutter urinals. On ground level, there are two restrooms; the men's restroom has 5 close-coupled toilets, 5 sinks and 4 individual ceramic urinals. The women's restroom has 6 sinks and 4 close-coupled toilets.

The weekly average population at the time of this study was 2405 students, according to UNICAMP's administration. There was also a fixed population of 11. The irrigated area of the yard is 30 m².

A survey was made with regards to the frequency of use of the toilets, cleaning of the floor and yard irrigation. To estimate the demand, as in the residential building, the existence of double activation toilets was supposed. In this survey, 125 students were interviewed.

The average number of daily activations as surveyed was 0.90 per student. During 4 weeks (20 days), the demand for February was estimated. For 31-day months a 1.107143 correction factor was applied and for 30-day months, a 1.071429 factor was applied.

Using this data, different scenarios were made to analyze the use of pluvial water, according to Table 4.

Scenario	Projected Uses	Volume (m³)		
		February	31 day Months	30 day Months
BD	Only flushing (double activation)	36.8	40.75	38.43
R	Only irrigation of the yard	0.86	0.96	0.93
L	Only floor washing	21.38	23.52	22.87
BD+R	Flushing (double activation) and irrigation of the yard	37.66	41.70	39.36
BD+L	Flushing (double activation) and floor washing	58.18	62.26	61.30
L+R	Irrigation of the yard and floor washing	22.24	24.47	23.80
BD+R+L	Flushing (double activation) and irrigation of the yard and floor washing	59.04	65.22	62.23

Table 4. Rainwater Demand for each scenario.

Scenarios	I	II	III	IV	V	VI	VII	VIII
BD	6.58	13.95			28.54	6.00	61.95	8.50
R	0.00	0.03			0.68	0.00	1.36	0.70
L	0.00	2.44			16.65	0.00	35.07	7.00
BD+R	8.48	13.35	256.28	91.34	29.22	8.00	61.61	8.50
BD+L	54.40	58.95			45.2	31.00	95.30	15.00
R+L	0.00	2.71			17.32	0.00	36.53	10.00
BD+R+L	61.25	70.08			45.87	32.00	96.72	16.00

NOTES:– Rippl Monthly; II – Rippl Daily Data; III – Azevedo Neto Pratical Method; IV – English Pratical Method; V – German Pratical Method; VI – Australian Pratical Method; VII – Weibull; VIII Netuno

Table 5. Reservation volumes obtained with the standard methods and Netuno software – Case Study 2 – institutional building.

As in case study 1, the economical evaluation was made by calculating the NPV of various methods, considering as cost only the construction of the storage.

In this case, there are various rainwater usage scenarios for the building. Hence, a series of variables have to be considered, such as the usage of the pluvial water and the utilized material (as in the previous case, 20 year lasting concrete and 10 year lasting fiber glass storages were analyzed). The same adjustments as in the previous case were utilized here as well: minimum, average and maximum in the period from 2001 to 2009, which were respectively: 5.59%, 10.88% and 19.63%. To illustrate, Fig.3 presents the NPV of fiberglass storage considering the minimum adjustment.

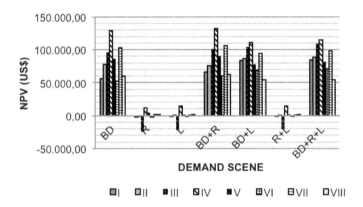

Fig. 3. Fiberglass storage NPV – minimum adjustment rate. Case study 2: Institutional Building.

It can be seen that the best NPV is yielded by the Brazilian practical method, utilizing fiberglass storages with cost higher than US$150,60.

Furthermore, scenarios with lower demand (R; L and R+L) are less favorable than others, nevertheless, if fiberglass is used, they show positive NPV for certain calculated volumes (practical Brazilian, English and German). The scenarios constituted for case study 2, independently of the method, were viable. This viability is largely due to high water taxes for this topology.

Another important factor is the large harvesting area of this building. This allows for a storage volume big enough to supply large demands, such as the ones estimated.

The result's analysis poses other questions such as:

- The Brazilian practical method usually yields large volumes. Despite this fact that when utilizing fiberglass, the results are economically interesting when compared to other methods.
- Results obtained using Weibull or Netuno methods are usually economically viable (NPV>0), independently of the demand scenario.
- The Practical English method yields high NPV values for the calculated storages, despite the material used, be it concrete or fiberglass.
- The lowest NPV values for volumes calculated were obtained using the Rippl Method, using either monthly or daily data.

The volume that yielded the highest NPV was calculated using the practical English method, for flushing and yard irrigation (BD+R), with 91.34 m³, with value higher than US$180,723, for the average adjustment rate.

If fiberglass storage were used, the highest NPV would be obtained with the volume calculated using the practical English Method in a scenario of demand, considering the average adjustment rate, approximately US$180,723.

As input to the simulation with Rain Toolbox, the total area is 1500 m² and the harvesting area is 1500 m². The storage height was limited to 3.00 m and its area to 5% of the total area. The simulation with 10 particles with 10 iterations yielded the results seen in Table 6.

Scenario	Concrete storage		Glass fiber storage	
	Volume (m³)	NPV (US$)	Volume (m³)	NPV (US$)
BD	161.24	223616.46	160.25	55835.69
L	1.00	6014.72	1.00	-1246.03
R	83.66	82617.96	7302	18840.31
BD+L	170.28	235312.23	16980	58903.62
BD+R	298.24	499238.69	295.38	128134.95
L+R	92.55	90275.09	75.62	20850.07
BD+L+R	303.39	511214.44	303.32	131277.15

Table 6. Reservation volumes obtained with Rain Toolbox. Case study 2 – Institutional building.

Analyzing the obtained results, it is possible to choose the highest NPV scenario, if the budget is large enough. Moreover, it can be seen that, economically, concrete storages are more advantageous than fiberglass storages.

Also, in scenario L, the financial return is very small or non-existent (if fiber glass is used).

The values for the storages calculated are relatively larger than the ones yielded by traditional methods. However, the utilization of these volumes can maximize financial return, making the implementation of rainwater harvesting systems more attractive.

4.2.3 Case 3 – Office building

An office building with 56 business rooms divided equally on two floors was selected for this case study. Each room has one close-coupled toilet and one washbasin. This building has not been built. The area in question is 1,431.40m², the garden is 675.65 m², the impermeable area is 2,942.15 m² and the internal area is about 443.45 m².

The estimated population is 757 people. The demand of rainwater was estimated based on the literature. The estimated total number of toilet flushes per person per day is 3 (2 flushes with approximately 3.5 L/f and 1 flush with about 7 L/f). An indicator of 1 L/m² was considered for landscape irrigation and for floor washing. The frequency of these activities is twice a week and once a week, respectively (Campos *et al.* 2003).

Based on these hypotheses, the rainwater demand for February was estimated as the demand pattern. For months with 31 and 30 days, correction factors of 1.107143 and 1.071429, respectively, were used. Table 7 shows de results. Table 8 shows the volumes obtained with use of those methods.

Scenario	Projected Uses	Volume (m³)		
		February	31 day Months	30 day Months
BD	Only flushing (double activation)	205,90	227,96	220,61
R	Only irrigation of the yard	8,12	8,99	8,70
L	Only floor washing	13,54	14,99	14,51
BD+R	Flushing (double activation) and irrigation of the yard	214,02	236,95	229,31
BD+L	Flushing (double activation) and floor washing	219,45	242,97	235,12
L+R	Irrigation of the yard and floor washing	21,66	23,98	23,21
BD+R+L	Flushing (double activation) and irrigation of the yard and floor washing	227,56	251,94	243,82

Table 7. Rainwater demand for different scenarios of use – case study 3 – office building

Rainwater demand scenarios	Volume of the reservoir (m³)							
	I	II	III	IV	V	VI	VII	VIII
BD	1038.6	1087.3			115.6	186.0	334.8	10.5
R	0.0	0.5			6.3	0.0	13.2	4.0
L	0.0	1.1			10.6	0.0	21.9	5.0
BD+R	1118.4	1167.8	300.2	107.0	115.6	195.0	348.0	10.5
BD+L	1171.6	1222.1			115.6	200.0	356.6	10.0
L+R	0.0	2.1			10.6	0.0	21.9	5.0
BD+R+L	1251.3	1305.9			115.6	209.0	369.8	10.5

Table 8. Reservation volumes obtained with the standard methods and Netuno software – Case Study 3 – Office Building.

The NPV was determined for 6 situations: lifetime of 10 years (fiberglass tanks) and 20 years (concrete Tanks) and 3 readjustment rates of water tariff, based on historical data: minimum, average and maximum. Figure 4 shows the results for the average readjustment rate.

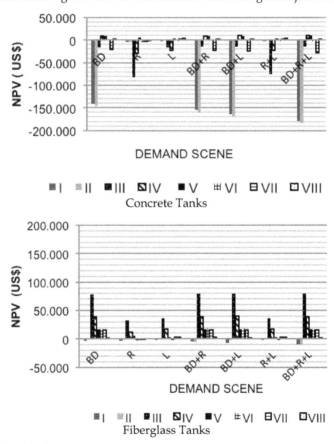

Note: US$ 1.00 = R$ 1.66 (02/18/2011)

Fig. 4. NPV for concrete/fiberglass tanks - average readjustment

All input data were shown earlier. Additionally, the following data was considered: Height of 3,00 m for the reservoir, percentage of the lot will be occupied by the reservoir: 5% of the total area of the lot and simulation with 10 particles and 10 interactions. Table 9 shows the results. The commercial opportunities of the use of the simulation are related to investments that can be considered infeasible or not so feasible, which could discourage investments in rainwater harvesting systems.

Raiwater demand scenarios	Concrete Tanks		Fiberglass Tanks	
	Vol(m³)	NPV(US$)	Vol(m³)	NPV (US$)
BD	101.4	1351650.72	101.4	329909.77
L	5.0	14560.97	5.0	-1705.45
R	51.8	37620.16	46.2	4338.77
BD+L	101.4	1354093.47	101.4	329909.77
BD+R	101.4	1389173. 78	101.4	337771.93
L+R	51.9	37620.15	45.2	4338.87
BD+L+R	101.4	1389173.78	101.4	337771.93

Table 9. Volumes and NPV using Rain Toolbox®

Besides that, the method proposed a factor that was not considered elsewhere. Economic variables are also important to stimulate the use of alternative sources of water, mainly for non-potable uses.

4.2.4 Case 4 – Commercial building (industrial plant)

The fourth and last case is a building in an industrial complex in the city of Paulinia, located only 5 km from the other cases analyzed in this work. This building is comprised of 4 pavements, in the first there is a kitchen and a refectory, in the other the administrative offices of the complex can be found.

Each pavement has two men's and two women's restrooms. On the ground level, aside from the four restrooms, there are two changing rooms, one for each gender. The kitchen has a capacity for 250 meals/day and a total of 180 workers.

The covered area is 291.40 m². The building has a 410.55 m² garden and an impermeable area of 677.13 m².

Similarly to cases 2 and 3, rainwater demand scenarios were made (BD, R, L, BD+R, BD+L; L+R e BD+L+R). Taking into account that the building was not constructed yet, the consumption data and usage of the sanitary facilities of the consulted bibliography were estimated.

Thus, 3 flushes/day*person were projected (Tomaz, 2000), 2 with partial volume and 1 with the total volume. One L/m² for the garden's irrigation was estimated, three times a week; and 1 L/m² to wash the floors, once a week. Considering 4 weeks (28 days), the demand for February was estimated. For 31-day months a 1.107143 correction factor was applied and for 30-day months, a 1.071429 factor was applied. Table 10 shows the results yielded.

The reservation volumes determined by the different methods are presented in Table 11.

Scenario	Volume (m³)		
	February	31 day Months	30 day Months
BD	48.96	54.20	52.46
R	4,96	5.45	5.28
L	2.71	3.00	2.90
BD+R	53.89	59.66	57.74
BD+L	51.69	57.20	55.36
L+R	7.63	8.45	8.18
BD+R+L	56.59	62.66	60.64

Table 10. Rainwater demand for the considered scenarios

Rainwater demand scenarios	Volume of the reservoir (m³)							
	I	II	III	IV	V	VI	VII	VIII
BD	295.76	308.15			22.22	5.00	78.75	5.00
R	0.00	0.70			3.21	0.00	7.92	4.50
L	0.00	0.22			1.77	0.00	4.35	4.00
BD+R	351.56	364.13	61.11	21.78	22.22	7.00	86.61	5.00
BD+L	325.09	337.88			22.22	5.00	83.04	5.00
L+R	1.02	2.67			4.95	0.00	12.27	4.50
BD+L+R	384,50	398.44			22.22	7.00	90.96	5.00

Table 11. Rainwater demand for different scenarios of use – case study 4 – office building industrial plant

Similarly to the previous cases, the economical analysis was carried out by calculating each scenario's NPV. The previously used adjustment rates are used here as well. Fig. 5 presents the results yielded using the average adjustment rate.

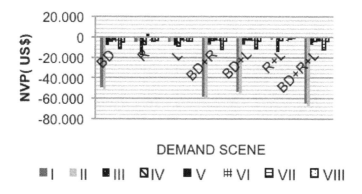

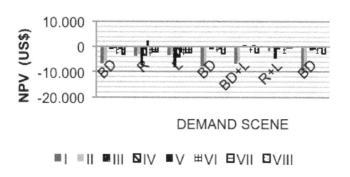

Fig. 5. NPV for concrete/fiberglass tanks - average readjustment

Even considering the maximum adjustment rate of the historical series, most scenarios remain unviable, with negative NPV.

In the case of concrete storages, only the volume determined using the Practical German Method for the L scenario and the Practical Brazilian Method for BD+R, BD+L and BD+R+L yielded positive NPV. The highest value, however, was calculated using the volume found with the Practical German Method for the R scenario (US$7,721.08).

For fiberglass storages, aside from the aforementioned scenarios, the NPV positive values were yielded by the Rippl method be it with daily or monthly data, for the BD, BD+R, BD+L and BD+R+L. The highest NPV was found using the volume determined with the Rippl method, with daily data for the BD+L scenario, which was US$7,687.34.

Given the results, for case study 4 only the irrigation scenario would be viable (NPV>0) if the storage used had 3.21 m³ of volume, value yielded by the Practical German Method. Furthermore, considering average and minimum adjustment scenarios, which are more realistic, this case has a positive NPV.

This is unviable largely due to the small harvesting area in relation to the relatively high demand, which calls for larger volumes.

Furthermore, not only in this case but also in others, even if the largest NPV volumes were to be utilized, one cannot be sure that it would yield the best results.

Considering this and maintaining the same input data as in the previous case studies (maximum storage height of 3.00m, maximum area of 5% of the total land area and the simulation with 10 particles and 10 iterations), the following NPV values were calculated for each volume and presented in Table 12.

Scenario	Concrete storage		Glass fiber storage	
	Volume (m³)	NPV (US$)	Volume (m³)	NPV (US$)
BD	163.15	137349,4	52.99	24293,02
L	5.00	12019,8	5.00	-2405,91
R	5.00	12019,8	5.00	-2405,91
BD+L	163.15	143905	49.02	25368,73
BD+R	160.40	146723,1	45,13	25943,1
L+R	5.00	12019,8	5,00	-2405,91
BD+L+R	131.43	151674,7	44.99	26586,67

Table 12. Volumes and NPV using Rain Toolbox®

4.2.5 Comparative analysis

Tables 13 and 14 show the best results yielded by the sensibility analysis and the model proposed in this work, respectively for concrete and fiberglass storages.

Case		Best result (scenario)	
study		Sensibility Analysis	Rain Toolbox
1	Volume (m³)	1.00	3.00
	NPV (US$)	-2970	-891.86
2	Volume (m³)	91.3 (BD+R)	303.3 (BD+R+L)
	NPV (US$)	191775.02(BD+R)	511214.4 (BD+R+L)
3	Volume (m³)	107.00 (R)	101.4 (BD+R+L)
	NPV (US$)	10678.55 (R)	1337301 (BD+R+L)
4	Volume (m³)	3.21 (R)	131.4 (BD+R+L)
	NPV (US$)	3091.67 (R)	151674.70 (BD+R+L)

Table 13. Best results yielded by sensibility analysis and by Rain Toolbox - concrete storage.

Case		Best Result (scenario)	
Study		Sensibility Analysis	Rain Toolbox
1	Volume (m³)	1.00	3.00
	NPV (US$)	-2470,28	-5795,67
2	Volume (m³)	91.3 (BD+R+L)	303.3 (BD+R+L)
	NPV (US$)	157052.76 (BD+R+L)	131277.20 (BD+R+L)
3	Volume (m³)	300.2 (BD)	101.4 (BD+R+L)
	NPV (US$)	79475.76 (BD)	339806.70 (BD+R+L)
4	Volume (m³)	3.24 (R)	45.00 (BD+R+L)
	NPV (US$)	2534.17 (R)	26586.67 (BD+R+L)

Table 14. Best results yielded by sensibility analysis and by Rain Toolbox - concrete storage.

It can be seen that the use of economic criteria to size storages is an interesting alternative that solves the lack of criteria in determining the volume. Moreover, the use of sensibility analysis, though extremely laborious, yields economically satisfactory results. The use of PSO as a way to incorporate was also very effective, providing the decision maker another investment opportunity, seeking the best possible return.

Analyzing with software, it is observed that the gain from the use of the volumes determined by the proposed method for cases 3 and 4 is evident: not only was the highest NPV found, but the demand also was completely supplied. For cases 1 and 2, the yield by the sensibility analysis is larger than the ones yielded by the proposed method. This is due to the fact that different adjustment factors were used in each method. Even though the minimum, average and maximum values were used in the sensibility analysis, the results selected for comparative analysis were the ones corresponding to an average adjustment rate.

Some of the volumes determined using the Rain Toolbox can be considered high, but they are limited by available land, never occupying more than 5% of its total free area.

With this method of sizing reservoirs, it is possible to make investments in rainwater harvesting systems more attractive, as there is a possibility of financial return.

This is only one way to think about the sizing of these system's reservoirs. Evidently a hydrological analysis of the system must be performed, but it has to be noted that the system is part of a building, increasing its costs, and they must frequently be viable not only environmentally, but also economically and financially.

The method proposed also seeks to solve a common problem in other such methods, which is the incompatibility of the storage's volume and land availability. This is the case especially in urban areas, where there this is a problem with other methods, which take the proposed method into account, fixing a maximum percentage of the land's area for the storage to occupy. The development of the computational tool contributes to facilitate the implantation of these concepts, incorporating a more fitting sizing method, considering the aforementioned aspects.

5. Conclusion

This article's main objective was to evaluate the incorporation of economical factors and land occupation for the dimensioning of rainwater harvesting system storages.

For this purpose, two methods were analyzed: firstly, sensibility analysis of various demand, water tariff adjustment and storage service life scenarios. Secondly the use of PSO as optimization technique of the NPV function, yielding the volume that gives the highest NPV value, considering a maximum limit of land occupation.

Both methods are viable to determine the reservation volume, however PSO revealed itself as the more interesting alternative, since the developed software will enable the decision of whether the system should be implemented and the optimal volume and it can reveal previously dismissed opportunities.

This technique's biggest advantage is its flexibility. It is possible, at certain moments, to introduce new variables to help determine the storage's volume, and it works well with one or multiple variables. Other limiting factors could be included in proposed method, such as initial investment, which allows this software to yield a volume compatible with the investor's budget. On the other hand, it is considered that future studies may clarify aspects not touched upon in this work, such as the inclusion of further parameters that can interfere with the decision-making and the behavior of the system in different rainfall patterns, as enhancements.

It is our hope that this work will effectively contribute to the enhancement of storages, increasing the number of these systems, improving conservation of water in buildings and helping urban draining.

6. Abbreviation list

GA – Genetic Algorithms
Gbest – Global best
NPV – Net Present Value
Pbest – personal best
PSO – Particle Swarm Optimization

7. References

Caraciolo, M.; Fernandes, D.; Bockholt, T.& Soares, L.Artificial Intelligence In Motion. (2010). Artificial Intelligence in Motion, In: *http://http://aimotion.blogspot.com*. Date of access January, 21st 2010, Available from: <http://aimotion.blogspot.com>

Associação Brasileira De Normas Tecnicas. Nbr 15527: Água De Chuva – Aproveitamento de coberturas em áreas urbanas para fins não potáveis – Requisitos.. Rio de Janeiro. Sept. 2007

Boeringer. D. W.& Werner. D.H (2004).Particle Swarm Optimization Versus Genetic
    Algorithms for Phased Array Synthesis. *IEEE Transactions on Antennas and
    Propagation,* Vol. 3, 3, (March, 2004), pp. (771-779), ISSN 0018-926X
Campos. M. A. S. (2004)Aproveitamento de água pluvial em edfifícios residenciais
    multifamiliares na cidade de São Carlos São Carlos. 2004. 131 f. MasterDegree
    Thesis - Federal University of Sao Carlos, Brazil.
Carrilho. O. J. B. (2007)Algoritmo Híbrido para Avaliação da Integridade Estrutural: Uma
    abordagem Heurística. São Carlos. 2007. 162 f. Doctoral Degree Thesis – São Carlos
    School of Engineering. University of Sao Paulo, Brazil.
Fewkes,A. (1999).The use of rainwater for WC flushing: the field-testing of a collection
    system. *Building and Environment,* Vol. 34, 6, (November, 1999), pp. (765-772), ISSN
    0360-1323
Ghisi, E.; Bressan, D.L. & Martini, M. (2007). Rainwater tank capacity and potential for
    potable water savings by using Rainwater in the residential sector of southeastern
    Brazil. *Building and Environment,* Vol. 42, 4, (April, 2007), pp. (1654-1666), ISSN
    0360-1323
Gould, J. & Nissen-Pettersen, E. (1999). *Rainwater catchment systems for domestic supply.* (2nd
    edition), Intermediate TEchonology Publications. ISBN 1-85339-456-4, United
    Kingdom
Group Raindrops. (2002).*Aproveitamento de água de chuva.* (1st edition), Organic
    Trading.ISBN85-87755-02-1, Brazil.
Liao. M.C.; Cheng. C. L.; Liu. Y. C.& Ding. J.W. (2005). Sustainable approach of existing
    building rainwater system from drainage to harvesting in Taiwan. *Proceedingsof CIB
    W062 Symposium on Water Supply and Drainage in Buildings,* Brussels, Belgium,
    September of 2005.
Liaw, C. H. & Tsai. Y.L. (2004). Optimal Storage Volume of rooftop rainwater harvesting
    systems for domestic use. *Journal of the American Water Resources Association,* Vol. 40,
    4, (August, 2004), pp. (901-912), ISSN 1093-474X
Montalvo, I. J.; Izquierdo. S.; Schwarze R. &Pérez-García. (2010). Multi-objective Particle
    Swarm Optimization applied to water distribution systems design: an approach
    with human interaction.*Proceedingsof International Congress on Environmental
    Modelling and Software,* Ottawa, Canada, July of 2010.
Rocha. V. L. (2009). Avaliação do potencial de economia de água potável e
    dimensionamento de reservatórios de sistemas de aproveitamento de água pluvial
    em edificações. Master Degree Thesis - Federal University of Santa Catarina, Brazil.
Simioni. W. I.; Ghisi. E & Gómez. L. A. (2004). Potencial de economia de água tratada
    através do aproveitamento de águas pluviais em postos de combustíveis :estudo de
    caso. *Proceedingsof Conferência Latino-Americana de Construção Sustentável/Encontro
    Nacional de Tecnologia do Ambiente Cosntruído,* , São Paulo, Brazil, July of 2004.
Thomaz, P. (2003). *Aproveitamento de água de chuva: aproveitamento de água de chuva para Áreas
    urbanas e fins não potáveis.* (1st edition), Navegar Editora. ISBN 85-87678-26-4, Brasil.
Yang. K. & Zhai. J. (2009). Particle Swarm optimization Algorithms for Optimal scheduling
    of Supply systems. *Proceedingsof International Symposium on Computational
    Intelligence and Design,* China, December of 2009.

Yruska. I.; Braga. L.G & Santos, C.. (2004). Viability of precipitation frequency use for reservoir sizing in Condominiums. *Journal of Urban and Environmental Engineering,* Vol. 4, 1, (January, 2010), pp. (23-28), ISSN 1982-3922

Wang. Y-G.. Kuhnert. P.. Henderson. B. & Stewart. L (2009). Reporting credible estimates of river loads with uncertainties in Great Barrier Reef catchments.. *Proceedingsof International Congress on Modeling and Simulation,* Australia, ISBN 978-0-9758400-7-8. July of 2009.

Permissions

The contributors of this book come from diverse backgrounds, making this book a truly international effort. This book will bring forth new frontiers with its revolutionizing research information and detailed analysis of the nascent developments around the world.

We would like to thank Manoj K. Jha, Ph.D., for lending his expertise to make the book truly unique. He has played a crucial role in the development of this book. Without his invaluable contribution this book wouldn't have been possible. He has made vital efforts to compile up to date information on the varied aspects of this subject to make this book a valuable addition to the collection of many professionals and students.

This book was conceptualized with the vision of imparting up-to-date information and advanced data in this field. To ensure the same, a matchless editorial board was set up. Every individual on the board went through rigorous rounds of assessment to prove their worth. After which they invested a large part of their time researching and compiling the most relevant data for our readers. Conferences and sessions were held from time to time between the editorial board and the contributing authors to present the data in the most comprehensible form. The editorial team has worked tirelessly to provide valuable and valid information to help people across the globe.

Every chapter published in this book has been scrutinized by our experts. Their significance has been extensively debated. The topics covered herein carry significant findings which will fuel the growth of the discipline. They may even be implemented as practical applications or may be referred to as a beginning point for another development. Chapters in this book were first published by InTech; hereby published with permission under the Creative Commons Attribution License or equivalent.

The editorial board has been involved in producing this book since its inception. They have spent rigorous hours researching and exploring the diverse topics which have resulted in the successful publishing of this book. They have passed on their knowledge of decades through this book. To expedite this challenging task, the publisher supported the team at every step. A small team of assistant editors was also appointed to further simplify the editing procedure and attain best results for the readers.

Our editorial team has been hand-picked from every corner of the world. Their multi-ethnicity adds dynamic inputs to the discussions which result in innovative outcomes. These outcomes are then further discussed with the researchers and contributors who give their valuable feedback and opinion regarding the same. The feedback is then collaborated with the researches and they are edited in a comprehensive manner to aid the understanding of the subject.

Apart from the editorial board, the designing team has also invested a significant amount of their time in understanding the subject and creating the most relevant covers. They scrutinized every image to scout for the most suitable representation of the subject and create an appropriate cover for the book.

The publishing team has been involved in this book since its early stages. They were actively engaged in every process, be it collecting the data, connecting with the contributors or procuring relevant information. The team has been an ardent support to the editorial, designing and production team. Their endless efforts to recruit the best for this project, has resulted in the accomplishment of this book. They are a veteran in the field of academics and their pool of knowledge is as vast as their experience in printing. Their expertise and guidance has proved useful at every step. Their uncompromising quality standards have made this book an exceptional effort. Their encouragement from time to time has been an inspiration for everyone.

The publisher and the editorial board hope that this book will prove to be a valuable piece of knowledge for researchers, students, practitioners and scholars across the globe.

List of Contributors

Marcelo Marcel Cordova and Enedir Ghisi
Federal University of Santa Catarina, Department of Civil Engineering, Laboratory of Energy Efficiency in Buildings, Florianópolis – SC, Brazil

Albert Rango and Kris Havstad
USDA-ARS Jornada Experimental Range, Las Cruces, New Mexico, USA

Atsushi Tsunkeawa, Mitsuru Tsubo and Derege Meshesha
Arid Land Research Center, Tottori University, Hamasaka, Japan

Abraha Gebrekiros and Eyasu Yazew
Department of Land Resources Management and Environmental Protection, Mekelle University, Ethiopia

Nigussie Haregeweyn
Arid Land Research Center, Tottori University, Hamasaka, Japan
Department of Land Resources Management and Environmental Protection, Mekelle University, Ethiopia

Shrikant Daji Limaye
UNESCO-IUGS-IGCP Project 523 "GROWNET", International Association of Hydrogeologists (IAH)
Ground Water Institute, India
Association of Geoscientists for International Development (AGID), UK
International River Foundation, Brisbane, Australia

Fardin Sadegh-Zadeh, Samsuri Abd Wahid, Bahi J. Seh-Bardan and Alagie Bah
Department of Land Management, Faculty of Agriculture, Universiti Putra Malaysia, Serdang, Selangor, Malaysia

Espitman J. Seh-Bardan
Department of Water Science, Faculty of Agriculture, Zabol University, Iran

John P. Hoehn
Michigan State University, United States of America

Petia Mijaylova Nacheva
Mexican Institute of Water Technology, Mexico

Marina Sangoi de Oliveira Ilha
Department of Architecture and Construction, School of Civil Engineering, Architecture and Urban Design, University of Campinas, Campinas, SP, Brazil

Marcus André Siqueira Campos
School of Civil Engineering, Federal University of Goiás, Goiânia, GO, Brazil

Printed in the USA
CPSIA information can be obtained
at www.ICGtesting.com
JSHW011341221024
72173JS00003B/190